中国地质大学(武汉)实验教学系列教材
中国地质大学(武汉)实验技术研究经费资助出版
中国地质大学(武汉)一般教学改革项目资助

数据库实验教程

何珍文　罗　晖　张夏林　编著

中国地质大学出版社
ZHONGGUO DIZHI DAXUE CHUBANSHE

内容简介

数据库是计算机科学与技术的重要基础之一,是计算机科学与技术、地理信息系统、空间信息与数字技术等信息相关专业的核心课程。本书是数据库课程的实验教程,通过数据库建库、管理及应用开发等一系列实验,帮助学生进一步理解所学数据库知识,提升运用数据库技术解决实际问题的能力。本书包括数据库管理系统软件、数据库与数据表操作、视图创建与使用、数据库的安全性与完整性、触发器、数据库设计与应用开发、数据库备份与恢复等方面的内容。

本书可以作为高等学校计算机科学与技术、地理信息系统、空间信息与数字技术、软件工程、通信工程、测绘、遥感及其他相关专业开设的数据库课程的本科生和研究生实验教材,也可供相关科研、企事业单位的研究与开发人员参考和使用。

图书在版编目(CIP)数据

数据库实验教程/何珍文,罗晖,张夏林编著.—武汉:中国地质大学出版社,2019.10

ISBN 978-7-5625-4641-2

Ⅰ.①数…

Ⅱ.①何… ②罗… ③张…

Ⅲ.①数据库系统-高等学校-教材

Ⅳ.①TP311.13

中国版本图书馆 CIP 数据核字(2019)第 230756 号

数据库实验教程	何珍文 罗 晖 张夏林 **编著**
责任编辑:彭 琳	责任校对:张咏梅
出版发行:中国地质大学出版社(武汉市洪山区鲁磨路388号)	邮政编码:430074
电 话:(027)67883511　　传 真:(027)67883580	E-mail:cbb@cug.edu.cn
经 销:全国新华书店	http://cugp.cug.edu.cn
开本:787毫米×1092毫米 1/16	字数:359千字　　印张:14
版次:2019年10月第1版	印次:2019年10月第1次印刷
印刷:武汉市籍缘印刷厂	印数:1—1000册
ISBN 978-7-5625-4641-2	定价:58.00元

如有印装质量问题请与印刷厂联系调换

中国地质大学(武汉)实验教学系列教材编委会名单

主　任: 刘勇胜

副主任: 徐四平　殷坤龙

编委会成员: (按姓氏笔画排序)

文国军　朱红涛　祁士华　毕克成　刘良辉

阮一帆　肖建忠　陈　刚　张冬梅　吴　柯

杨　喆　金　星　周　俊　章军锋　龚　健

梁　志　董元兴　程永进　窦　斌　潘　雄

选题策划:

毕克成　李国昌　张晓红　赵颖弘　王凤林

《数据库实验教程》

作者名单

何珍文　罗　晖　张夏林　张军强

刘　刚　田宜平　张志庭　李新川

孙亚博　龙仕容　赵　洪　陈　奔

前　言

在当今大数据时代，数据应用已经渗透到各行各业，成为重要的生产因素。随着信息技术的不断发展，世界上每时每刻都在不断产生着海量的数据。所谓数据库，就是长期存储在计算机内、有组织、可共享的大量数据的集合。数据库技术为这些数据的存储、管理、查询访问提供了有效的解决方案，广泛应用于企业管理、电子金融、电子商务、电子政务、空间信息系统等各个方面，成为信息技术发展的重要支撑。

在多年教学科研基础上，我们编写了这本数据库实验教材，供本科生和研究生在实验中使用。本书注重理论联系实际，课程实验软件平台涉及 Oracle、SQL Server 等主流商业数据库管理系统软件，同时也兼顾了一些开源软件或项目，如 MySQL、SQLite 等；通过任务和问题设计，将数据库的一般原理与实际操作相结合，强化学生的实际操作，提高开发数据库及其应用程序来解决实际问题的能力。

本书分为两个部分。第一部分为数据库实验部分，主要包括数据库管理系统软件、数据库与数据表操作、数据表的数据操作、视图的创建与使用、数据库安全性、数据库完整性、触发器、数据库设计、存储过程和函数、数据库应用开发(C++/Java/C#)、数据备份与恢复以及综合实验。该部分内容按照数据库管理系统的操作使用、数据库设计、数据库应用开发、综合实验顺序展开，逐渐深入解析，加深了学生们对数据库基本概念的理解，强化了学生对基本知识的掌握和对基本技术的运用，提升了他们分析问题和解决问题的能力。第二部分为参考答案，与第一部分相对应，提供了 Oracle、MySQL 和 SQL Server 三种主流数据库管理系统的参考答案。

本书第 1 章由何珍文编写；第 2~4 章由何珍文、罗晖、王嫒妮编写；第 5~14 章由何珍文、张军强、张夏林编写；张军强编写了 MySQL 答案并进行了核校；罗晖对 SQL Server 部分答案进行了核校；孙亚博、龙仕容、赵洪等研究生参与了本书的部分内容编写、图形编绘以及代码编写与调试工作。课程组刘刚、张夏林、田宜平、翁正平、李章林等老师为本书提出了许多的宝贵建议。全书由何珍文统稿并定稿。

数据库技术发展日新月异，特别是大数据技术的发展对数据库提出了许多新的要求，带来了诸多挑战。受时间、篇幅、知识面和材料限制，本书还存在很多不足之处，期待您的批评指导，以便我们在后续工作中不断完善相关内容。

何珍文
2019 年 3 月

目 录

第一部分 数据库实验

1 数据库管理系统软件 ·· (3)
　1.1 Oracle 数据库管理系统 ··· (3)
　1.2 MySQL 数据库管理系统 ··· (13)
　1.3 SQL Server 数据库管理系统 ·· (20)
　1.4 其他数据库管理系统 ·· (25)

2 数据库和数据表操作 ··· (26)
　2.1 实验目的 ··· (26)
　2.2 实验平台 ··· (26)
　2.3 实验内容 ··· (26)
　2.4 实验报告 ··· (33)

3 数据表的数据操作 ·· (34)
　3.1 实验目的 ··· (34)
　3.2 实验平台 ··· (34)
　3.3 实验内容 ··· (34)
　3.4 实验报告 ··· (43)

4 视图的创建与使用 ·· (44)
　4.1 实验目的 ··· (44)
　4.2 实验平台 ··· (44)
　4.3 实验内容 ··· (44)
　4.4 实验报告 ··· (48)

5 数据库安全性 ··· (49)
　5.1 实验目的 ··· (49)
　5.2 实验平台 ··· (49)
　5.3 实验内容 ··· (49)

  5.4 实验报告 ……………………………………………………………………… (52)

6 数据库完整性 ………………………………………………………………… (53)
  6.1 实验目的 ……………………………………………………………………… (53)
  6.2 实验平台 ……………………………………………………………………… (53)
  6.3 实验内容 ……………………………………………………………………… (53)
  6.4 实验报告 ……………………………………………………………………… (56)

7 触发器 …………………………………………………………………………… (57)
  7.1 实验目的 ……………………………………………………………………… (57)
  7.2 实验平台 ……………………………………………………………………… (57)
  7.3 实验内容 ……………………………………………………………………… (57)
  7.4 实验报告 ……………………………………………………………………… (62)

8 数据库设计 ……………………………………………………………………… (63)
  8.1 实验目的 ……………………………………………………………………… (63)
  8.2 实验平台 ……………………………………………………………………… (63)
  8.3 实验内容 ……………………………………………………………………… (63)
  8.4 实验报告 ……………………………………………………………………… (68)

9 存储过程与函数 ………………………………………………………………… (69)
  9.1 实验目的 ……………………………………………………………………… (69)
  9.2 实验平台 ……………………………………………………………………… (69)
  9.3 实验内容 ……………………………………………………………………… (69)
  9.4 实验报告 ……………………………………………………………………… (74)

10 数据库应用开发(C++) ……………………………………………………… (75)
  10.1 实验目的 …………………………………………………………………… (76)
  10.2 实验平台 …………………………………………………………………… (76)
  10.3 实验内容 …………………………………………………………………… (76)
  10.4 实验报告 …………………………………………………………………… (101)

11 数据库应用开发(Java) ……………………………………………………… (102)
  11.1 实验目的 …………………………………………………………………… (102)
  11.2 实验平台 …………………………………………………………………… (103)
  11.3 实验内容 …………………………………………………………………… (103)
  11.4 实验报告 …………………………………………………………………… (108)

12 数据库应用开发(C♯) ……………………………………………………… (109)
  12.1 实验目的 …………………………………………………………………… (111)

  12.2 实验平台 ……………………………………………………………………………… (111)
  12.3 实验内容 ……………………………………………………………………………… (111)
  12.4 实验报告 ……………………………………………………………………………… (115)

13 数据备份与恢复 …………………………………………………………………………… (116)
  13.1 实验目的 ……………………………………………………………………………… (116)
  13.2 实验平台 ……………………………………………………………………………… (116)
  13.3 实验内容 ……………………………………………………………………………… (116)
  13.4 实验报告 ……………………………………………………………………………… (119)

14 综合实验(课程设计) ………………………………………………………………………… (120)

附录1 实验报告要求 …………………………………………………………………………… (121)

附录2 课程设计报告要求 ……………………………………………………………………… (123)

第二部分 参考答案

1 MySQL 的参考答案 ……………………………………………………………………… (127)
  数据库管理系统软件参考答案 ………………………………………………………… (127)
  数据库和数据表操作参考答案 ………………………………………………………… (127)
  数据表的数据操作参考答案 …………………………………………………………… (129)
  视图的创建与使用参考答案 …………………………………………………………… (134)
  数据库安全性参考答案 ………………………………………………………………… (135)
  数据库完整性参考答案 ………………………………………………………………… (137)
  触发器参考答案 ………………………………………………………………………… (139)
  数据库设计参考答案 …………………………………………………………………… (142)
  存储过程与函数参考答案 ……………………………………………………………… (146)
  数据库应用开发(C++)参考答案 ……………………………………………………… (150)
  数据库应用开发(Java)参考答案 ……………………………………………………… (153)
  数据库应用开发(C♯)参考答案 ……………………………………………………… (157)
  数据备份与恢复参考答案 ……………………………………………………………… (161)
  综合实验(课程设计)参考答案(略) …………………………………………………… (162)

2 SQL Server 的参考答案 ………………………………………………………………… (163)
  数据库管理系统软件参考答案 ………………………………………………………… (163)
  数据库和数据表操作参考答案 ………………………………………………………… (163)
  数据表的数据操作参考答案 …………………………………………………………… (165)
  视图的创建与使用参考答案 …………………………………………………………… (170)

数据库安全性参考答案 …………………………………………………… (171)

数据库完整性参考答案 …………………………………………………… (173)

触发器参考答案 …………………………………………………………… (176)

数据库设计参考答案 ……………………………………………………… (180)

存储过程与函数参考答案 ………………………………………………… (185)

数据库应用开发(C++)参考答案 ………………………………………… (189)

数据库应用开发(Java)参考答案 ………………………………………… (201)

数据库应用开发(C#)参考答案 …………………………………………… (202)

数据备份与恢复参考答案 ………………………………………………… (205)

综合实验(课程设计)参考答案(略) ……………………………………… (210)

主要参考文献 …………………………………………………………… (211)

第一部分

数据库实验

1 数据库管理系统软件

本章主要介绍了三种常用的数据库管理系统软件:Oracle、SQL Server 和 MySQL。通过安装和初步使用可初步了解这几种常见的数据库管理系统。

1.1 Oracle 数据库管理系统

本节主要介绍 Oracle 数据库管理系统的发展历史、主要特征,然后介绍了 Oracle 数据库管理系统的安装。

1.1.1 Oracle 数据库简介

Oracle 公司(甲骨文公司)成立于 1977 年,是全球最大的信息管理软件及服务供应商。其主要软件产品 Oracle 数据库系统软件已经成为市场占有率最高的数据库产品。最早的 Oracle 是采用汇编语言在 DEC 公司的 PDP-11 上开发完成的。Oracle 3 则采用了 C 语言开发,使得 Oracle 具有了较强的跨平台优势。1997 年,Oracle 公司推出了基于 Java 的 Oracle 8,两年后推出了 Oracle 8i,其中的 i 则代表了 Internet,并添加了 SQLJ 和 XML 等特性。

2001 年,Oracle 公司发布了 Oracle 9i。这个版本最重要的一个新特性就是增加了 RAC。2003 年,Oracle 公司又发布了 Oracle 10g,其中的 g 代表了 Grid(网格)。这个版本最重要的特性就是加入了网格计算功能。2007 年,Oracle 公司发布了 Oracle 11g。2013 年,Oracle 公司发布了 Oracle 12c。从 2017 年 7 月开始,Oracle 公司改变了以往的数据库软件发布流程,采用年度发布和季度更新的策略。Oracle 公司新策略中的发布频率明显加快,能够更快地推出新功能。目前最新的版本是 Oracle 19c。对于数据库基本原理的学习与实验而言,Oracle 11g 及其后续版本都可以选取。由于本书需要安装多种数据库管理系统,为了使得虚拟机尽可能小且方便实习使用,本书以空间占用较少,同时支持 64 位和 32 位的 Oracle 11g 版本为例进行讲解。

1.1.2 Oracle 数据管理系统的安装

Oracle 数据库系统是一个大型的商业数据库系统软件,如由 Oracle 公司授权和安装光盘,可以直接在安装光盘中双击"SETUP.EXE"开始执行安装,也可以在 Oracle 的官方网站 http://www.oracle.com/technetwork/database/enterprise-edition/downloads/index.html 下载。

下载后,双击"SETUP.EXE"程序,开始 Oracle 11g 的安装向导,如图 1-1 所示。在图

1-1的界面中可以填入一个电子邮箱用于接受Oracle发送的安全配置信息。如果不想要这些信息,可以直接点击"下一步(N)",系统会弹出如图1-2所示的错误提示,直接点击"是(Y)"按钮,进入下一步安装。

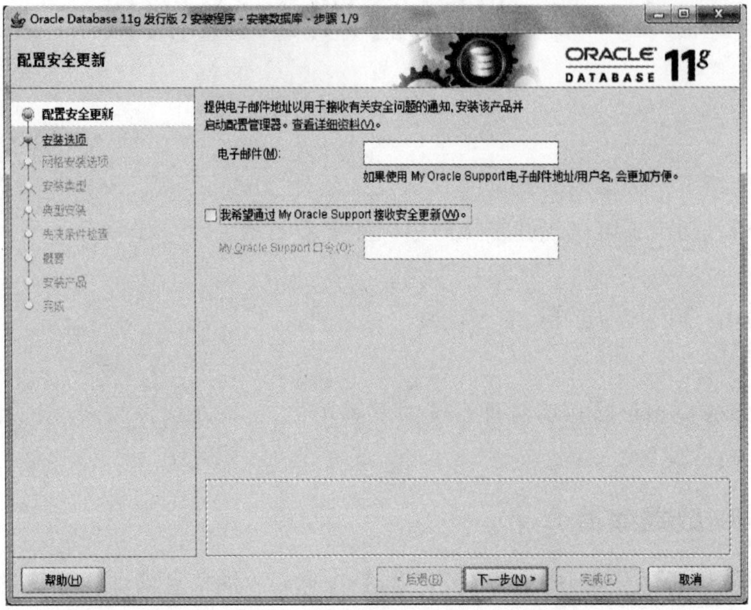

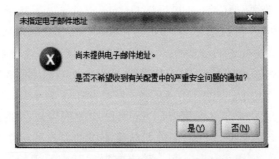

图1-1 Oracle 11g安装的安全更新配置

图1-2 安全配置中的电子邮件错误

在图1-3中显示了三种安装选项,分别是"创建和配置数据库(C)""仅安装数据库软件(I)""升级现有的数据库(U)"。如果只想安装一个数据库管理系统软件,而不需要安装程序自动创建一个数据实例,可以选择"仅安装数据库软件(I)";如果在安装的目标机器上已经有一个Oracle数据库,则可以选择"升级现有的数据库(U)"。在这里我们是全新安装,并且希望安装程序自动创建数据库实例,所以应该选取"创建和配置数据库(C)"。

接下来会出现如图1-4所示的界面。在这个界面中可以选择安装的系统类型,如桌面类和服务器类。由于我们的数据库管理系统软件只是安装在普通的PC上面供实习使用,因此应该选择桌面类的系统,因为服务器类的系统一般对计算机的软硬件资源要求相对较高,并且对资源独占性较强,不适合作为实习安装选择。

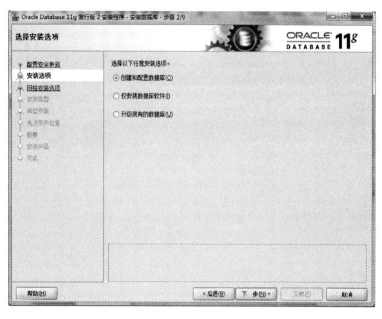

图1-3 Oracle 11g 的安装选项

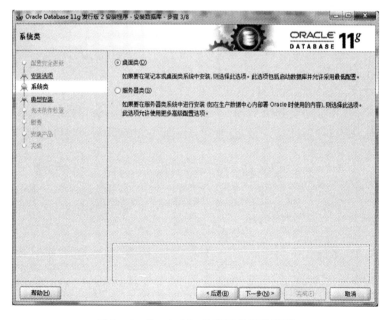

图1-4 Oracle 11g 的安装的系统类型

接下来会出现如图1-5所示的界面。在这个界面中主要填写 Oracle 的相关目录、字符集等信息。如我们填写的 Oracle 的基目录是 D:\Applications\Oracle\Administrator，则 Oracle 相应的软件位置会变成 D:\Applications\Oracle\Administrator\product\11.2.0\dbhome_1，而 Oracle 的数据文件位置会变成 D:\Applications\Oracle\Administrator\oradata。

图1-5 Oracle 11g 的安装配置

图1-5中除了有目录设置外,还有数据库版本,这里选择企业版本,相关的一些组件会自动安装。字符集可以选择默认的,它采用的是操作系统默认的字符集。全局数据库名比较重要,以后访问数据库的时候会经常用到,可以直接采用默认的 orcl,也可以自己命名。管理口令是管理员的初始口令,可以输入自己设定的口令,但是这个口令如果比较简单而不符合 Oracle的安全策略,安装程序会弹出如图1-6所示的提示对话框,提示密码不符合Oracle建议的标准,可以点击"否(N)"返回重新设定,也可以不管它继续执行下一步安装。

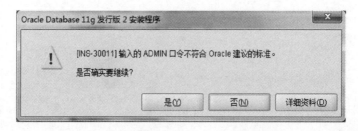

图1-6 Oracle 11g 的用户口令不合规范

接下来,安装系统会进行一些先决条件检查(图1-7),主要包括操作系统、CPU、内存等方面的检查。如果没有通过检查,则会弹出如图1-8所示的界面,它会提示哪些项没有通过安装程序的检查。对于配置比较低的机器,可能会出现内存不够大等问题,这些会在一定程度上影响 Oracle 数据库的性能,如果没有可以替换的机器而一定要在这台机器上进行Oracle安装,可以在图1-8中选择全部忽略,然后单击"下一步(N)",这样就会进行下一步安装,进入到如图1-9所示的界面。这个界面是前面所有安装配置信息的一个总结,它会提示即将要安装的数据库产品的一些概要,如数据的全局设置、产品清单信息、数据库信息等。如果确定这些信息是正确的,则可以单击"完成(F)"按钮,进入图 1-10 所示界面,开始 Oracle 11g 的具体产品的安装。

图 1-7　Oracle 11g 安装程序进行系统配置检查

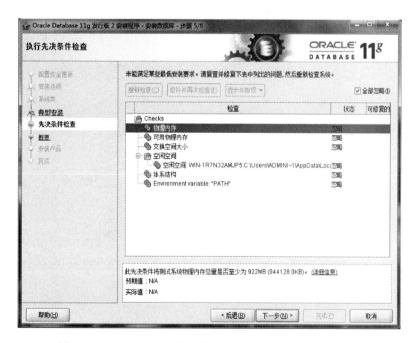

图 1-8　Oracle 11g 安装程序进行系统配置检查失败的处理

图1-9 Oracle 11g 安装概要

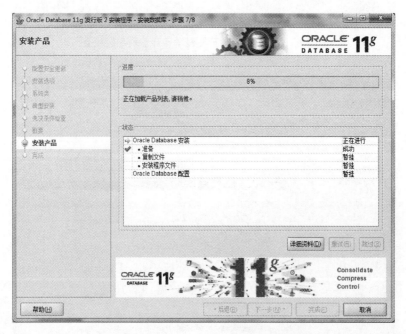

图1-10 Oracle 11g 安装具体产品过程

在安装过程中,由于安装程序要访问网络,因此,如果操作系统安装有防火墙,可能会弹出如图1-11所示的界面,最好点击"允许访问(A)",以避免通过网络连接访问 Oracle 服务器时被防火墙阻断。

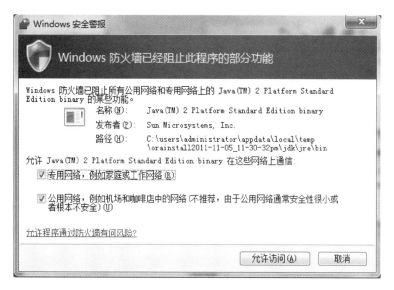

图 1-11　操作系统防火墙在 Oracle 11g 安装过程中出现的网络访问提示

接下来就是等待安装产品这一步完成，这是一个比较长的过程。完成后将显示如图 1-12 所示的界面，并开始数据库实例的创建和安装，如图 1-13 所示。这个步骤主要分为复制数据库文件、创建并启动 Oracle 数据库实例。同样这个过程耗时也比较长。数据库创建完成后，会弹出如图 1-14 所示的界面。在这个界面中可以看到安装的一些信息，如显示了安装路径、全局数据库名、系统标识符（SID）等数据库信息。

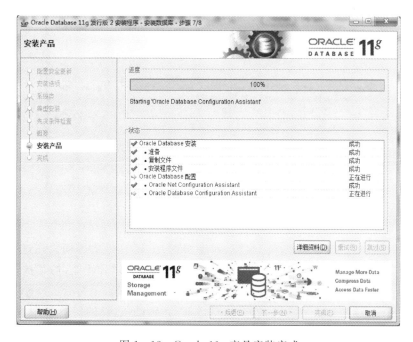

图 1-12　Oracle 11g 产品安装完成

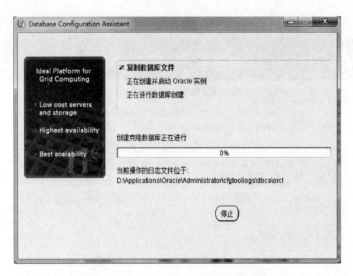

图 1-13　Oracle 11g 安装数据实例

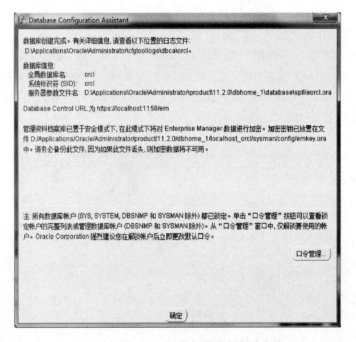

图 1-14　Oracle 11g 数据库实例创建完成

点击图 1-14 中的"口令管理…"按钮,弹出如图 1-15 所示的界面。在这个界面中显示了所有实例数据库用户的名称。这些用户中除了 SYS 和 SYSTEM 外,均是被默认锁定。我们可以为 SYS 和 SYSTEM 用户分别设置密码。同样,如果密码不符合 Oracle 建议的标准,安装程序会弹出一个如图 1-16 所示的提示对话框。Oracle 建议的口令复杂性是,口令的长度至少要 8 个字符,此外口令中至少应该有一个大写字符、一个小写字符和一个数字。当然,如果不愿意接受它的建议,可以直接选择"是"。这样就完成了 Oracle 数据库软件的安装,安装系统会弹出如图 1-17 所示的安装成功提示对话框。提示显示企业管理数据库控制台的访问 URL 如下：

Enterprise Manager Database Control URL -(orcl)：
https://localhost:1158/em

同时也会提醒数据库配置文件已经安装到 D:\Applications\Oracle\Administrator，相关组件也已经安装到 D:\Applications\Oracle\Administrator\product\11.2.0\dbhome_1。这样就完成了 Oracle 数据库软件安装的全部过程。

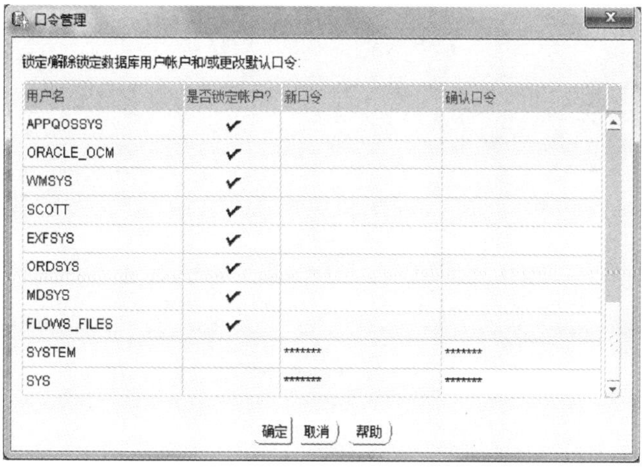

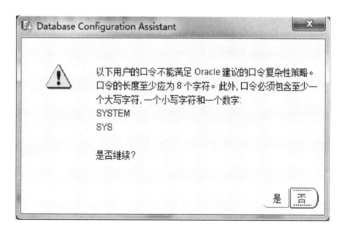

图 1-15　Oracle 11g 数据库实例的用户密码设定

图 1-16　密码强度不符合 Oracle 11g 规范

安装完成后会在操作系统的开始菜单上出现如图 1-18 所示的菜单项。我们可以点击"SQL Plus"来测试一下安装是否正确。点击"SQL Plus"菜单，弹出如图 1-19 所示的界面。在上面输入"sys as sysdba"，会提示输入用户密码，然后输入安装时设定的密码，会显示"SQL>"提示，则可以开始输入 SQL 语句"select * from v$version;"，就会在界面上显示查询结果，显示的是 Oracle 的版本信息。

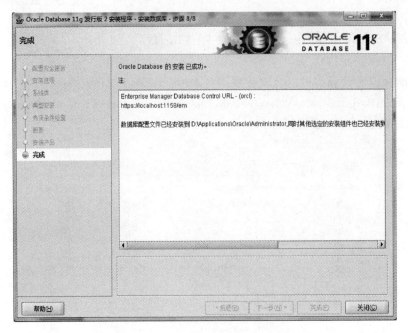

图1-17　Oracle 11g 整个安装过程完成

图1-18　Oracle 11g 整个安装过程完成

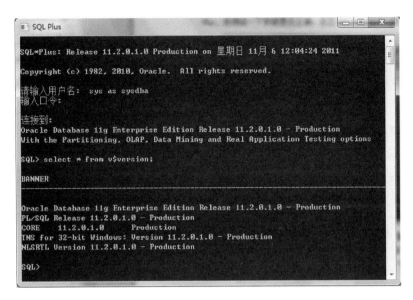

图 1-19　SQL Plus 启动界面

1.2　MySQL 数据库管理系统

1.2.1　MySQL 数据库简介

MySQL 是一个开源的中小型关系数据库管理系统，最初的开发者为瑞典 MySQL AB 公司。在 2008 年该公司被 Sun 公司收购。2009 年，Sun 公司又被 Oracle 公司收购。目前 MySQL 的开源社区项目由 Oracle 公司维护。MySQL 是一种关联数据库管理系统，关联数据库将数据保存在不同的表中，而不是将所有数据放在一个大仓库内，这样就增加了速度并提高了灵活性。MySQL 软件采用了 GPL(GNU 通用公共许可证)。由于其体积小、速度快、总体成本低，且具有开放源码这一特点，使许多中小型网站为了降低网站总体成本而选择了 MySQL 作为网站数据库。

MySQL 使用 C 和 C++编写，并使用了多种编译器进行测试，保证源代码的可移植性；支持 AIX、FreeBSD、HP-UX、Linux、Mac OS、Novell Netware、OpenBSD、OS/2 Wrap、Solaris、Windows 等多种操作系统；为多种编程语言，如 C、C++、Python、Java、Perl、PHP、Eiffel、Ruby 和 Tcl 等提供了 API；支持多线程，既能够作为一个单独的应用程序应用在客户端/服务器网络环境中，也能够作为一个库嵌入到其他的软件中提供多语言支持，如提供了 TCP/IP、ODBC 和 JDBC 等多种数据库连接途径；提供了用于管理、检查、优化数据库操作的管理工具；可以处理拥有上千万条记录的大型数据库。

同时 MySQL 支持多种存储引擎，MyISAM 是 MySQL 的默认存储引擎，拥有较高的插入、查询速度，但不支持事务；InnoDB 是 MySQL 的事务型数据库的首选存储引擎，支持 ACID 事务，支持行级锁定；另外还有来自 Berkeley DB 的 BDB，将所有数据置于内存的存储引擎、将一定数量的 MyISAM 数据表联合而成一个整体的 MERGE 存储引擎、Cluster/NDB 高冗余的存储引擎等。另外，MySQL 的存储引擎接口定义良好，用户可以按照接口开发自己的存储引擎。

与其他的大型数据库如 Oracle、DB2、SQL Server 等相比，MySQL 有它的不足之处，如规模小、功能有限（MySQL Cluster 的功能和效率都相对比较差）等，但是这丝毫也没有减少它受欢迎的程度。对于一般的个人使用者和中小型企业来说，MySQL 提供的功能已经绰绰有余，而且由于 MySQL 是开放源码软件，因此可以大大降低总体成本。目前 Internet 上流行的网站构架方式是 LAMP（Linux＋Apache＋MySQL＋PHP/Perl/Python）和 LNMP（Linux＋Nginx＋MySQL＋PHP/Perl/Python），即使用 Linux 作为操作系统，Apache 和 Nginx 作为 Web 服务器，MySQL 作为数据库，PHP/Perl/Python 作为服务器端脚本解释器。由于这四个软件都是免费或开放源码软件，因此使用这种方式不用花一分钱（除开人工成本）就可以建立起一个稳定、免费的网站系统。这也是目前 MySQL 成为非常流行的主流关系数据库管理系统之一的重要原因。

1.2.2　MySQL 数据库管理系统安装

MySQL 数据库系统是一个中小型的开源数据库系统软件，可以直接在 Oracle 公司的网站上下载相应的安装程序。我们以 Windows 32 位操作系统和 MySQL 5.5 为例（为便于构建小型虚拟机实习环境，选择 32 位操作系统和早期 MySQL 版本。读者可根据需要选取其他较新的 MySQL 数据库版本）。我们需要下载以下两个安装文件：

（1）mysql－5.5.16－win32.msi，MySQL 数据库服务器安装程序。

（2）mysql－workbench－gpl－5.2.34.2－win32.msi，MySQL 前端工具安装程序。

首先，单击"mysql－5.5.16－win32.msi"，出现如图 1－20 所示的界面。单击"Next"出现终端用户授权协议，开源的 MySQL 采用的是 GNU GENERAL PUBLIC LICENSE，如图 1－21 所示。

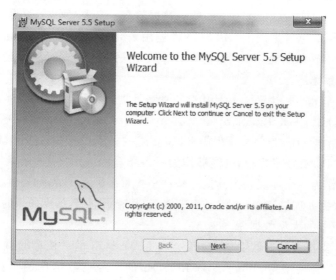

图 1－20　MySQL 安装启动界面

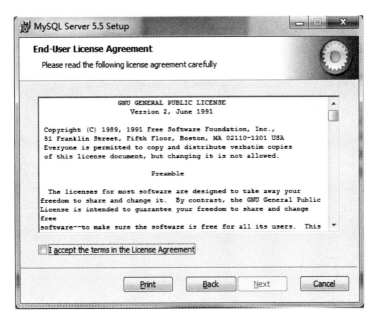

图 1-21　MySQL 的 GPL 授权协议

MySQL 服务器版本的安装类型有三种(图 1-22)：第一种是典型安装，将自动安装最常用的一些程序模块，推荐给大多数用户使用；第二种是用户自定义安装，允许用户选择哪些程序模块需要安装，并可以选择安装的位置，建议高级用户使用该模式；第三种是完全安装，就是所有的程序模块都将被安装，这需要最大的磁盘空间。由于 MySQL 本来就是一个中小型的数据库管理系统软件，如果空间允许，可以选择完全安装。

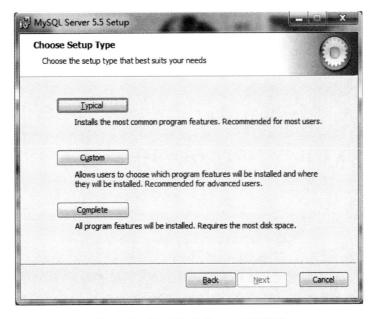

图 1-22　MySQL 的 Server 安装类型

在图 1-23 中单击"Install"按钮,开始 MySQL Server 的安装过程,然后在后续出现的界面上单击"Finish",完成 MySQL 服务器的安装,如图 1-24 所示。

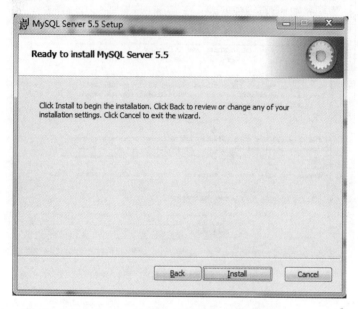

图 1-23　MySQL 的 Server 安装类型

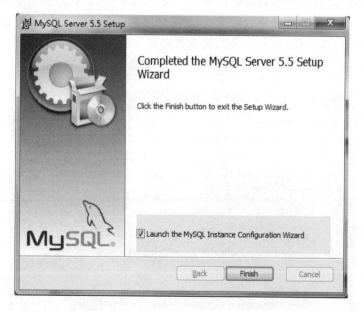

图 1-24　MySQL 的 Server 安装完成

MySQL 的 Server 安装完成后将开始 MySQL 实例的配置向导,用以引导用户进行本机 MySQL 服务实例的参数配置。图 1-25 显示的是 MySQL Server 实例配置类型,分为两种:一种是详细配置,可以为本机进行服务器的优化配置;另一种是标准配置,建议一般用户使用。

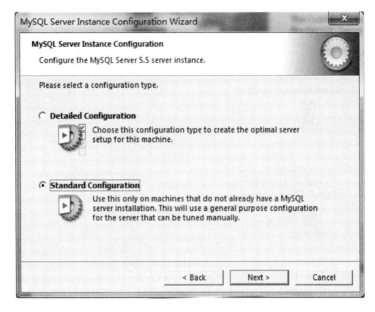

图 1-25　MySQL 的 Server 配置类型

图 1-26 显示的是 MySQL Server 的 Windows 服务属性配置,这里它的服务名为 MySQL,并且在 Windows 启动的时候自动启动,还将 MySQL 中的 bin 目录添加到了 Windows 的 Path 环境变量中。图 1-27 中显示的是 MySQL Server 实例的安全设置,这里用于设置 root 密码和是否创建匿名用户。

图 1-26　MySQL 的 Server 服务配置向导

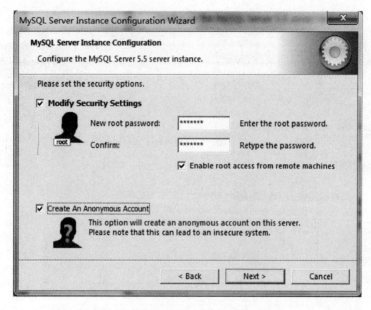

图 1-27 MySQL 的 Server 安全设置

MySQL 的 Server 实例配置完成后将出现如图 1-28 所示的提示界面,表示 MySQL 安装与配置完成。为了验证安装是否正确,可以在"开始"菜单中找到"MySQL 5.5 Command Line Clinet",点击出现如图 1-29 所示的界面,提示输入密码,输入正确的密码后出现 MySQL 的一些版本、版权及命令提示信息。在"mysql>"提示符后面输入"show databases",然后输入回车键,如果安装正确,将会显示本服务器上目前所有的数据库名称。至此我们就成功地安装了 MySQL 的数据库服务器程序。

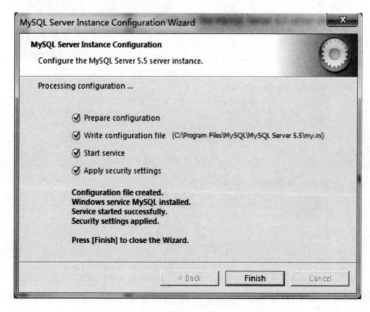

图 1-28 MySQL 的 Server 安装完成

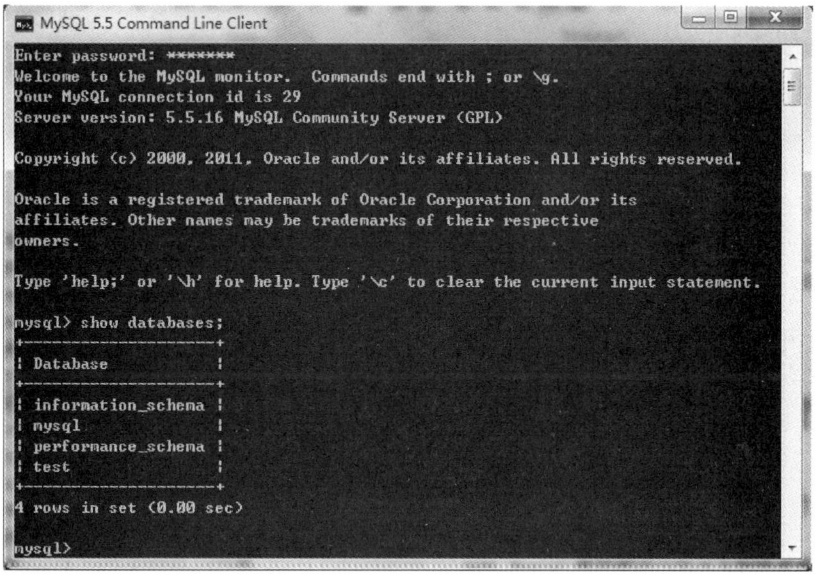

图 1-29 MySQL 的命令行客户端

接下来安装 Workbench,安装好后启动 Workbench,出现如图 1-30 所示的初始化界面。Workbench 是三个环境的集成,SQL Development 主要用于连接数据库并进行开发,Data Modeling 主要用于数据建模,Server Administration 则主要用于数据库服务器管理。具体的使用方法参见 Oracle 关于 Workbench 的帮助文档。

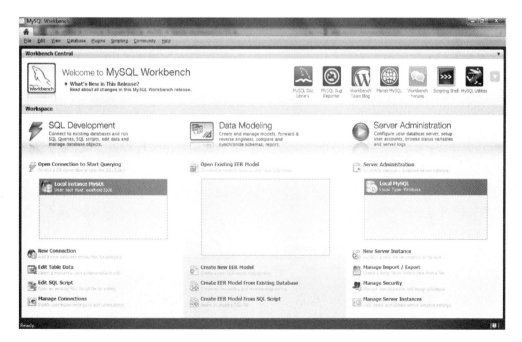

图 1-30 MySQL Workbench 启动界面

1.3 SQL Server 数据库管理系统

1.3.1 SQL Server 数据库简介

SQL Server 是目前流行的关系数据库管理系统之一。它最初是由 Microsoft 与 Sybase 等公司联合开发的。在微软的 Windows NT 操作系统推出后,微软将 SQL Server 移植到 Windows NT 系统上,专注于开发推广 SQL Server 的 Windows 系统的版本。比较具有里程碑意义的几个版本分别是 Microsoft SQL Server 2000、Microsoft SQL Server 2005、Microsoft SQL Server 2008 和 Microsoft SQL Server 2012。目前最新版本为 Microsoft SQL Server 2017。

Microsoft SQL Server 2000 是 Microsoft 公司推出的 SQL Server 数据库管理系统。该版本继承了 SQL Server 7.0 版本的优点,同时增加了许多更先进的功能,具有使用方便、可伸缩性好、与相关软件集成程度高等优点,可在 Windows 的多种平台上使用。

Microsoft SQL Server 2005 是一个全面的数据库平台,使用集成的商业智能(BI)工具提供了企业级的数据管理。Microsoft SQL Server 2005 数据库引擎为关系型数据和结构化数据提供了更安全可靠的存储功能,可以构建和管理用于业务的高可用和高性能的数据应用程序。Microsoft SQL Server 2005 不仅可以有效地执行大规模联机事务处理,而且可以完成数据仓库和电子商务应用等许多具有挑战性的工作。Microsoft SQL Server 2005 数据引擎是本企业数据管理解决方案的核心。此外,SQL Server 2005 结合了分析、报表、集成和通知功能,使企业可构建和部署经济有效的 BI 解决方案,帮助团队通过记分卡、Dashboard、Web services 和移动设备将数据应用推向业务的各个领域。与 Microsoft Visual Studio、Microsoft Office System 以及新的开发工具包(包括 Business Intelligence Development Studio)的紧密集成使 Microsoft SQL Server 2005 与众不同。

Microsoft SQL Server 2008 是一个重大的产品版本,它推出了许多新的特性和关键的改进,使得它成为非常强大、全面的 SQL Server 版本。与 Microsoft SQL Server 2005 相比,Microsoft SQL Server 2008 可以对整个数据库、数据文件和日志文件进行加密,而不需要改动应用程序,通过支持第三方密钥管理和硬件安全模块(HSM)产品为这个需求提供了很好的支持。同时,Microsoft SQL Server 2008 增强了审查功能,可使自己审查自己的数据操作,从而提高了遵从性和安全性。Microsoft SQL Server 2008 的一个更重要的改进是增加了对非关系数据的支持。

Microsoft SQL Server 2017 与 Microsoft SQL Server 2012 和 SQL Server 2008 相比,一个显著的增强就是云计算支持和跨操作系统支持。为了方便在不同的操作系统中安装实习系统,在本教材中采用 Microsoft SQL Server 2017 Express 进行教学。

1.3.2 SQL Server 数据库管理系统安装

SQL Server 的安装程序界面友好，提供了简单易用的安装向导，如图 1-31 所示。用户所要做的就是一路点击"下一步(N)"直到整个安装过程完成。图 1-32 显示了安装过程中版本的选择界面。Developer 版本免费，包含了 SQL Server 的完整功能，但只能用于开发和测试；Express 版本也免费，并可用于生产环境，但只能用于小规模使用场景。本书选择免费的 Express 版本。图 1-33 显示了 SQL Server 安装许可，选择"我接受许可条款(A)。"点击"下一步(N)"。接下来是 SQL Server 安装规则(图 1-34)，主要是数据库及操作系统方面的安全设置规则。再接下来是 SQL Server 功能选择(图 1-35)，可以全选，也可以选择部分，本书选取时排除了与机器学习相关的选择项。图 1-36 显示了 SQL Server 的实例配置，可以选择"默认实例(D)"，也可选择"命名实例(A)"，并给出实例名。SQL Server 的服务器提供了多个服务，如图 1-37 所示，采用默认的设置即可。在 SQL Server 的数据库引擎配置中，一般采用混合模式，并设置 sa 用户的密码，如图 1-38 所示。只要接受上述选择的设置，安装向导即可完成全部所选组件和数据库的安装与配置。

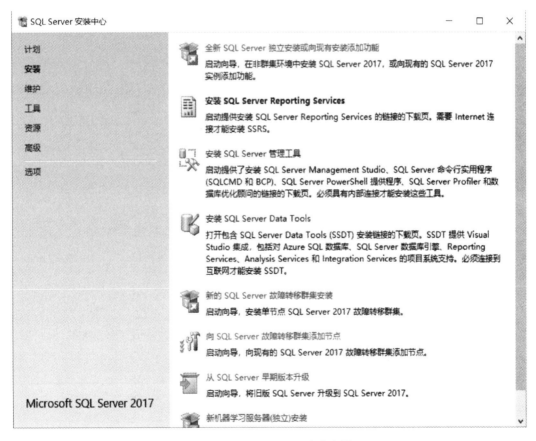

图 1-31 SQL Server 安装向导

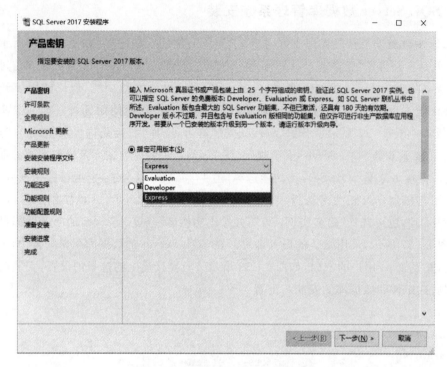

图 1-32　选择 SQL Server 安装版本

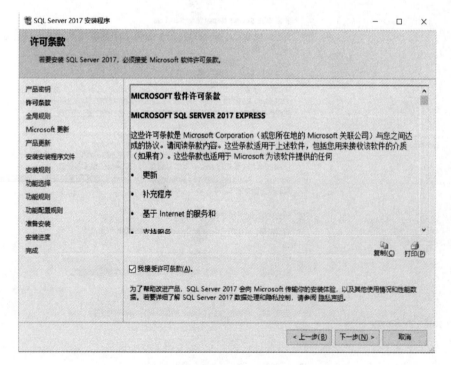

图 1-33　SQL Server 安装许可

图 1-34　SQL Server 安装规则

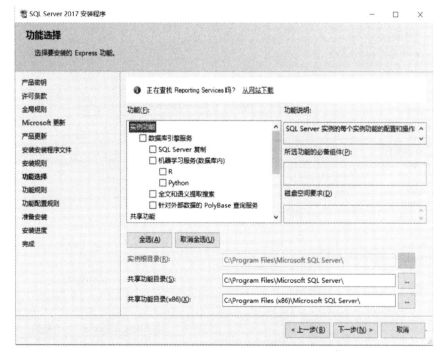

图 1-35　SQL Server 功能选择

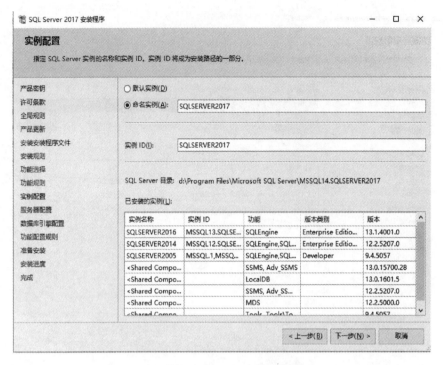

图 1-36　SQL Server 实例配置

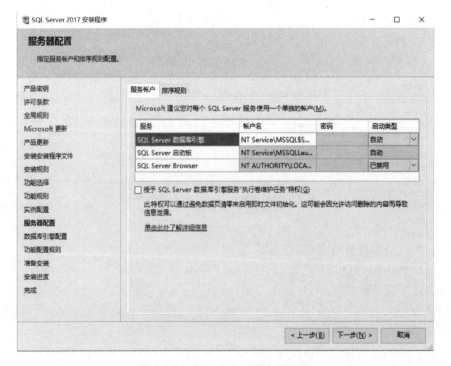

图 1-37　SQL Server 服务器配置

图 1-38 SQL Server 数据库引擎配置

1.4 其他数据库管理系统

除了上述三种主流的数据管理系统之外，还有一些比较著名的开源数据库系统，如大中型的 PostgreSQL 和小型的 SQLite 等。

PostgreSQL 是由加州大学伯克利分校开发的一款开源的数据库管理系统，支持大部分的 SQL 标准并且提供了很多其他现代特性，如复杂查询、外键、触发器、视图、事务完整性、多版本并发控制等。PostgreSQL 也可以通过增加新的数据类型、函数、操作符、聚集函数、索引方法、过程语言等方法进行扩展。该开源系统许可证较灵活，任何人都可以以任何目的免费使用、修改和分发 PostgreSQL。

SQLite 是一个实现自包含、无服务器、零配置、事务性的轻型开源数据库引擎。与大多数其他 SQL 数据库不同，SQLite 只是一个嵌入式 SQL 数据库引擎，没有单独的服务器进程。SQLite 直接读写普通磁盘文件，具有多个表、索引、触发器和视图的完整 SQL 数据库被包含在单个磁盘文件中，目前已经在很多嵌入式产品中推广使用。它的资源占用率非常低，能够支持 Windows/Linux/Unix/Android 等主流的操作系统，同时能够跟很多程序语言相结合，比如 C/C++、Java、C♯、PHP、Python 等。

2 数据库和数据表操作

2.1 实验目的

通过实习了解并掌握数据库和数据表的两种创建方式：
(1)通过数据库管理系统软件提供的图形化管理界面完成数据库和数据表的创建。
(2)通过 SQL 语言完成数据库和数据表的创建。

2.2 实验平台

(1)操作系统：Windows XP、Windows Server 2003 及后续版本、Windows 7 及后续版本。
(2)数据库管理系统：根据实际情况，自己选择 Oracle 或 SQL Server 或 MySQL 中的一种数据库管理系统软件。

2.3 实验内容

(1)建立一个数据库 UNIVERSITY,其中包括六个数据表。
(a)系的信息表 Department(<u>Dno</u>,Dname,Daddress)。
(b)学生信息表 Student(<u>Sno</u>,Sname,Ssex,Sage,Dno)。
(c)教师信息表 Teacher(<u>Tno</u>,Tname,Ttitle,Dno)。
(d)课程信息表 Course(<u>Cno</u>,Cname,Cpno,Ccredit)。
(e)学生选课表 SC(<u>Sno</u>,<u>Cno</u>,Grade)。
(f)教师授课表 TC(<u>Tno</u>,<u>Cno</u>,Site)。

上面加有下划线的代码为该表的关键码,Dno 表示系的编号,Dname 表示系名,Daddress 表示系所在的办公地址;Sno 表示学号,Sname 表示学生姓名,Ssex 表示学生性别,Sage 表示学生年龄;Tno 表示教师编号,即职工号,Tname 表示教师姓名,Ttitle 表示教师职称;Cno 表示课程编号,Cname 表示课程名称,Cpno 表示先导课程编号,Ccredit 表示课程学分;Grade 表示每个学生的每一门课的成绩;Site 表示授课地点。

第 1 步:启动 SQL Developer 或 Navicat,采用 sys 用户登录数据库,这里以 Oracle SQL Developer 软件为例,如图 2-1 所示。如果是 Oracle 12c 及其后续版本,并且希望在可插拔数据库中构建 UNIVERSITY 数据库,则以 sys 登录后,还需要执行 SQL 语句,打开可插拔数据

库实例 PDBORCL(默认安装),命令如下:

ALTER PLUGGABLE DATABASE PDBORCL OPEN;

或者如果希望打开所有的可插拔数据库实例,则可以运行下列 SQL 语句:

ALTER PLUGGABLE DATABASE ALL OPEN;

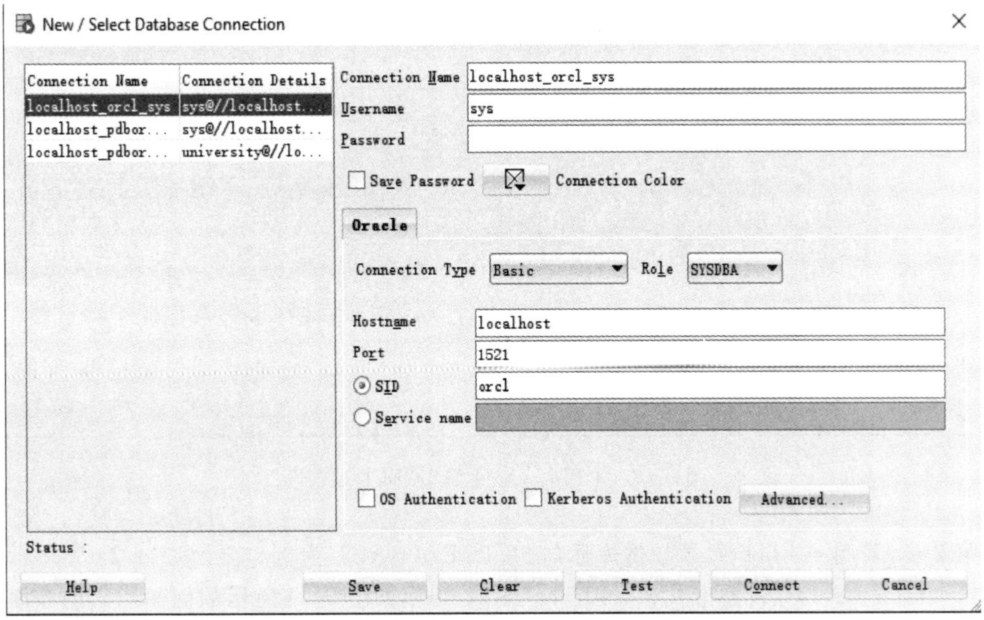

图 2-1 采用 sys 登录 ORCL 数据库

将 PDBORCL 数据库实例开启后,需要以 sys 或 system 登录连接到 PDBORCL,连接设置如图 2-2 所示。

这里需要注意的是,对于可插拔数据库实例一般采用服务名称的方式进行连接。如果 PDBORCL 服务名连接不通,则需要修改 Oracle 的 tnsnames.ora 文件。该文件一般位于{OracleHome}\NETWORK\ADMIN 目录下,在 tnsnames.ora 文件末尾添加:

```
PDBORCL=
  (DESCRIPTION=
    (ADDRESS=(PROTOCOL=TCP)(HOST=localhost)(PORT=1521))
    (CONNECT_DATA=
      (SERVER=DEDICATED)
      (SERVICE_NAME=pdborcl)
    )
  )
```

然后,重启 Oracle 的服务。如果采用的是 Oracle 11g 或之前的版本,则直接采用图 2-1 所示的方式登录即可。

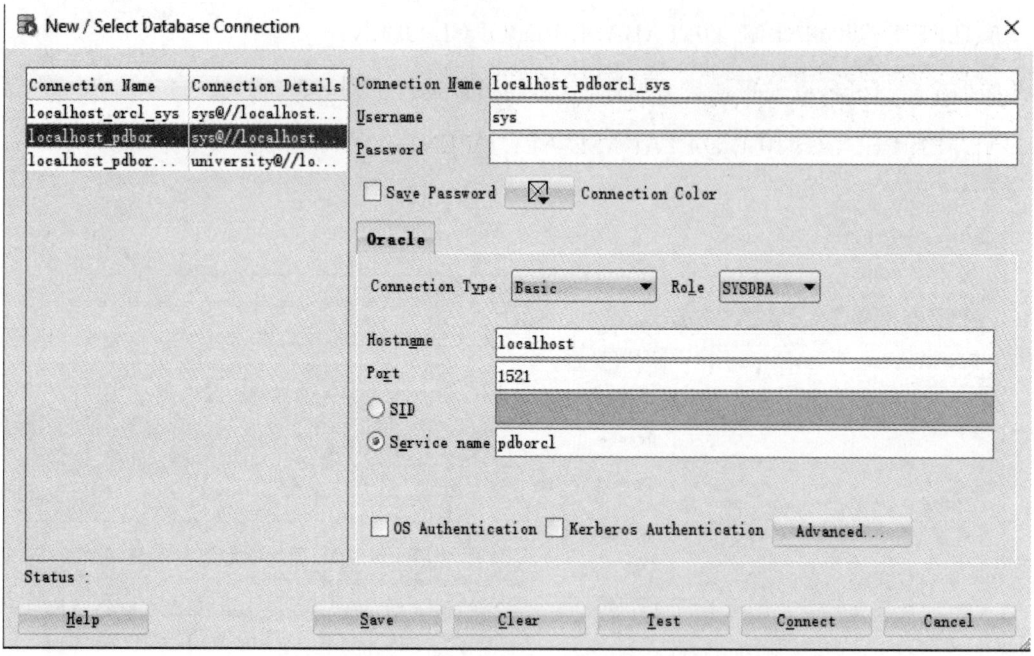

图 2-2 采用 sys 登录 PDBORCL 数据库

第 2 步：采用 sys 用户登录数据库后（这里以 PDBORCL 为例），界面如图 2-3 所示。右键单击"Other Users"，再点击"Create User"菜单，弹出如图 2-4 所示的创建用户的界面。在 User 选项卡中，填写用户信息，用户名称为 University，用户密码为 cug，默认表空间为 EXAMPLE 或 USERS，临时表空间为 TEMP。在 Granted Roles 选项卡中，勾选 CONNECT、DBA、RESOURCE 三项，如图 2-5 所示。点击"Apply"按钮，完成 University 用户创建。在 Oracle 数据库中，创建一个用户的时候会默认创建一个模式（Schema），并且与用户名同名。

第 3 步：接下来就可以采用 University 登录 Oracle 数据库，如图 2-6 所示。

第 4 步：进行数据表创建（图 2-7）。

(2) 采用 SQL 语言删除步骤 1 中建立的数据表和数据库。在 Oracle 中，一个数据库实例中会有很多个模式，要删除 UNIVERSITY 数据库的说法其实是不准确的，一般指的是要删除 UNIVERSITY 模式。可以直接采用删除 University 用户的方式来实现该功能：

DROP USER UNIVERSITY;

(3) 采用 SQL 语言建立数据库 UNIVERSITY。

CREATE USER UNIVERSITY IDENTIFIED BY "cug"
　　DEFAULT TABLESPACE "EXAMPLE"
　　TEMPORARY TABLESPACE "TEMP";
GRANT "DBA" TO UNIVERSITY;
GRANT "CONNECT" TO UNIVERSITY;
GRANT "RESOURCE" TO UNIVERSITY;

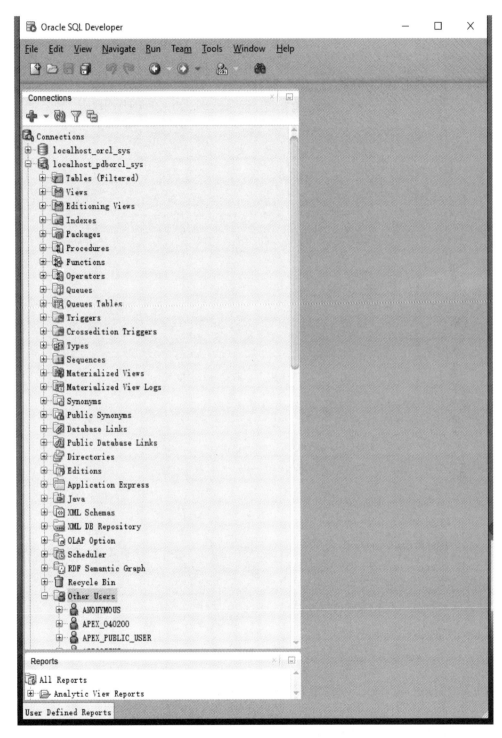

图 2-3 采用 sys 登录 PDBORCL 数据库后的主界面

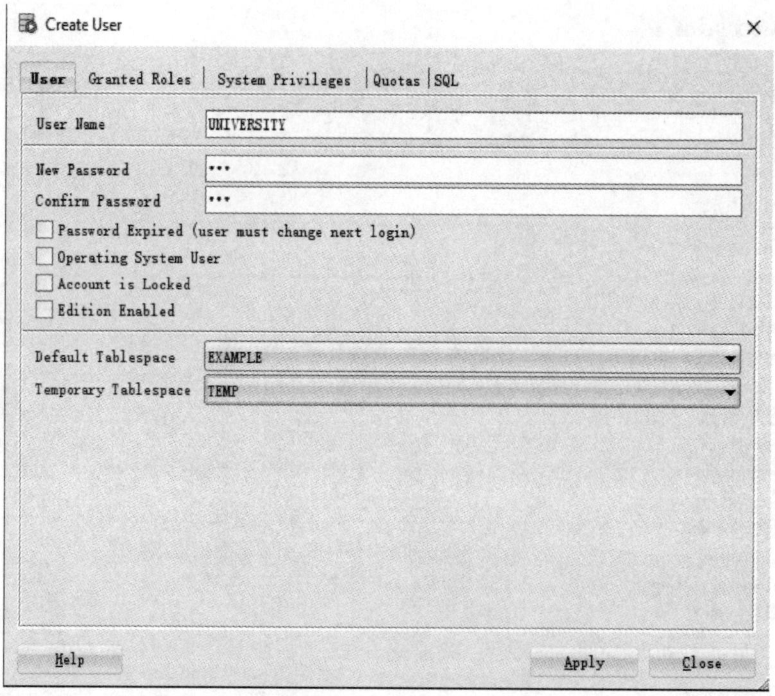

图 2-4　创建用户界面(User 选项卡)

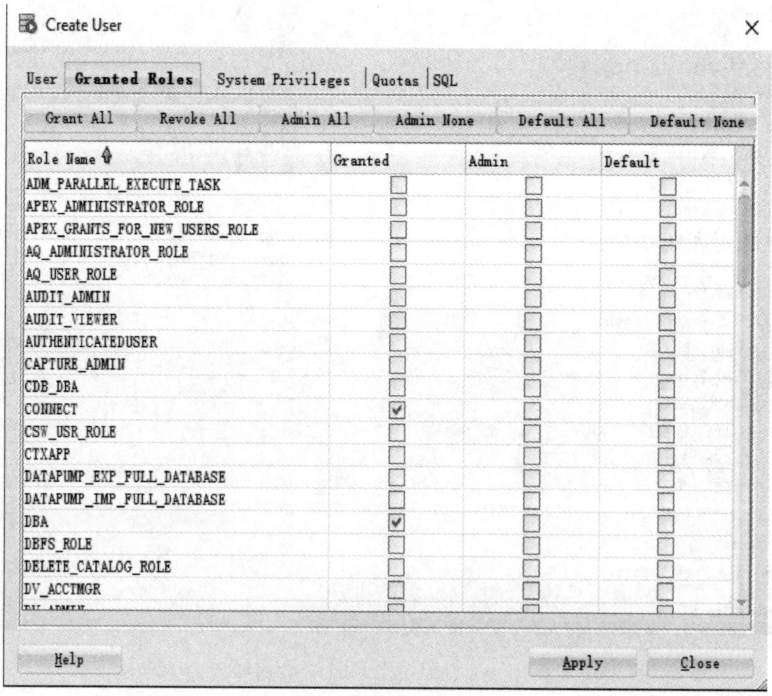

图 2-5　创建用户界面(Granted Roles 选项卡)

2　数据库和数据表操作

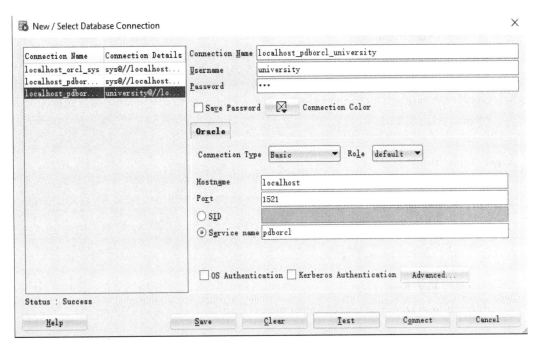

图 2-6　采用 University 用户连接数据库

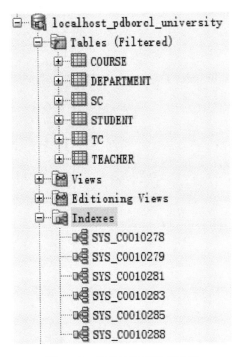

图 2-7　创建的数据表和索引

(4)采用 SQL 语言创建 UNIVERSITY 数据库中的六个数据表。

```sql
create table Department(
    Dno int,
    Dname varchar(50),
    Daddress varchar(50),
    primary key(Dno)
);
create table Student(
    Sno int,
    Sname varchar(50),
    Ssex varchar(2),
    Sage int,
    Dno int,
    primary key(Sno),
    foreign key(Dno)references Department(Dno)
);
create table Teacher(
    Tno int primary key,
    Tname varchar(50),
    Ttitle varchar(50),
    Dno int,
    foreign key(Dno)references Department(Dno)
);
create table Course(
    Cno int primary key,
    Cname varchar(50),
    Cpno int,
    CCredit int,
    foreign key(Cpno)references Course(Cno)
);
create table SC(
    Sno int,
    Cno int,
    Grade float,
    primary key(Sno,Cno),
    foreign key(Sno)references Student(Sno),
    foreign key(Cno)references Course(Cno)
);
```

```
create table TC(
    Tno int,
    Cno int,
    Site varchar(50),
    primary key(Tno,Cno),
    foreign key(Tno)references Teacher(Tno),
    foreign key(Cno)references Course(Cno)
);
```

(5)采用 SQL 语言为 Student 表的 Sname 建立唯一索引(图 2-8)。

`create unique index Sname_index on Student(Sname);`

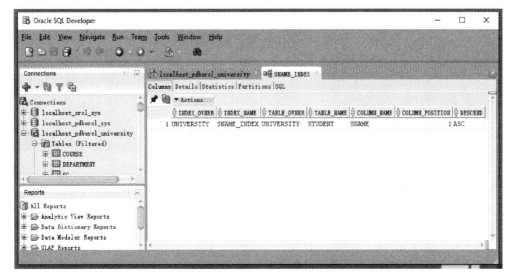

图 2-8　创建的 SNAME_INDEX

(6)采用 SQL 语言删除 Student 表 Sname 上的唯一索引。

`drop index Sname_index;`

(7)采用 SQL 语言给 Teacher 表添加一个字段 Tsex 教师性别。

`alter table Teacher add Tsex varchar(2);`

(8)采用 SQL 语言删除 Teacher 表中的字段 Tsex。

`alter table Teacher drop column Tsex;`

2.4　实验报告

实验报告按照附录 1 的格式进行编写。

3 数据表的数据操作

3.1 实验目的

通过实习了解并掌握数据表中的数据操作,包括向表中插入数据、修改数据、删除数据等。
(1)通过数据库管理系统软件提供的管理界面完成数据表的相关数据操作。
(2)通过 SQL 语言完成数据表的查询、插入、修改、删除等相关数据操作。

3.2 实验平台

(1)操作系统:Windows XP、Windows Server 2003 及后续版本、Windows 7 及后续版本。
(2)数据库管理系统:根据实际情况,自己选择 Oracle 或 SQL Server 或 MySQL 中的一种数据库管理系统软件。
(3)本实验以第 2 章构建的 UNIVERSITY 数据库为例进行实验。

3.3 实验内容

(1)分别采用 UI 界面和 SQL 语言为 UNIVERSITY 的 Department 表输入数据。数据显示结果如图 3-1 所示。
插入数据的 SQL 语句如下:

```
insert into Department(Dno,Dname,Daddress)values(1,'地球科学学院','主楼东');
insert into Department(Dno,Dname,Daddress)values(2,'资源学院','主楼西');
insert into Department(Dno,Dname,Daddress)values(3,'材化学院','材化楼');
insert into Department(Dno,Dname,Daddress)values(4,'环境学院','文华楼');
insert into Department(Dno,Dname,Daddress)values(5,'工程学院','水工楼');
insert into Department(Dno,Dname,Daddress)values(6,'地球物理与空间信息学院','物探楼');
insert into Department(Dno,Dname,Daddress)values(7,'机械与电子信息学院','教二楼');
insert into Department(Dno,Dname,Daddress)values(8,'经济管理学院','经管楼');
insert into Department(Dno,Dname,Daddress)values(9,'外语学院','北一楼');
insert into Department(Dno,Dname)values(10,'信息工程学院');
```

insert into Department(Dno,Dname,Daddress)values(11,'数学与物理学院','基委楼');
insert into Department(Dno,Dname,Daddress)values(12,'珠宝学院','珠宝楼');
insert into Department(Dno,Dname,Daddress)values(13,'政法学院','政法楼');
insert into Department(Dno,Dname,Daddress)values(14,'计算机学院','北一楼');
insert into Department(Dno,Dname)values(15,'远程与继续教育学院');
insert into Department(Dno,Dname)values(16,'国际教育学院');
insert into Department(Dno,Dname,Daddress)values(17,'体育部','体育馆');
insert into Department(Dno,Dname,Daddress)values(18,'艺术与传媒学院','艺传楼');
insert into Department(Dno,Dname,Daddress)values(19,'马克思主义学院','保卫楼');
insert into Department(Dno,Dname,Daddress)values(20,'江城学院','江城校区');

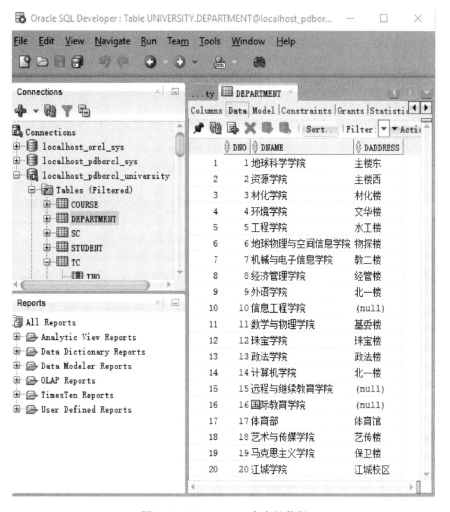

图 3-1 Department 表中的数据

(2)分别采用 UI 界面和 SQL 语言为 UNIVERSITY 的 Student 表输入数据。数据显示结果如图 3-2 所示。

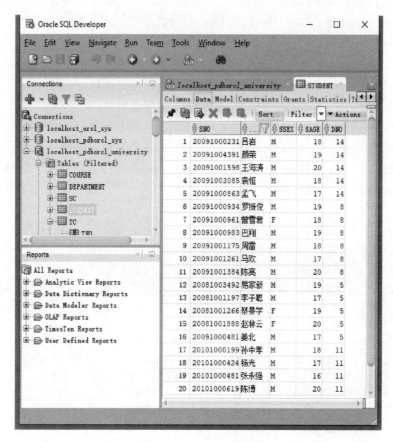

图3-2 Student表中的数据

插入数据的SQL语句如下：

insert into Student(Sno,Sname,Ssex,Sage,Dno) values(20091000231,'吕岩','M',18,14);
insert into Student(Sno,Sname,Ssex,Sage,Dno) values(20091004391,'颜荣','M',19,14);
insert into Student(Sno,Sname,Ssex,Sage,Dno) values(20091001598,'王海涛','M',20,14);
insert into Student(Sno,Sname,Ssex,Sage,Dno) values(20091003085,'袁恒','M',18,14);
insert into Student(Sno,Sname,Ssex,Sage,Dno) values(20091000863,'孟飞','M',17,14);
insert into Student(Sno,Sname,Ssex,Sage,Dno) values(20091000934,'罗振俊','M',19,8);
insert into Student(Sno,Sname,Ssex,Sage,Dno) values(20091000961,'曾雪君','F',18,8);
insert into Student(Sno,Sname,Ssex,Sage,Dno) values(20091000983,'巴翔','M',19,8);
insert into Student(Sno,Sname,Ssex,Sage,Dno) values(20091001175,'周雷','M',18,8);
insert into Student(Sno,Sname,Ssex,Sage,Dno) values(20091001261,'马欢','M',17,8);
insert into Student(Sno,Sname,Ssex,Sage,Dno) values(20091001384,'陈亮','M',20,8);
insert into Student(Sno,Sname,Ssex,Sage,Dno) values(20081003492,'易家新','M',19,5);
insert into Student(Sno,Sname,Ssex,Sage,Dno) values(20081001197,'李子聪','M',17,5);
insert into Student(Sno,Sname,Ssex,Sage,Dno) values(20081001266,'蔡景学','F',19,5);

insert into Student(Sno,Sname,Ssex,Sage,Dno)values(20081001888,'赵林云','F',20,5);
insert into Student(Sno,Sname,Ssex,Sage,Dno)values(20091000481,'姜北','M',17,5);
insert into Student(Sno,Sname,Ssex,Sage,Dno)values(20101000199,'孙中孝','M',18,11);
insert into Student(Sno,Sname,Ssex,Sage,Dno)values(20101000424,'杨光','M',17,11);
insert into Student(Sno,Sname,Ssex,Sage,Dno)values(20101000481,'张永强','M',16,11);
insert into Student(Sno,Sname,Ssex,Sage,Dno)values(20101000619,'陈博','M',20,11);
insert into Student(Sno,Sname,Ssex,Sage,Dno)values(20101000705,'汤文盼','M',18,11);
insert into Student(Sno,Sname,Ssex,Sage,Dno)values(20101000802,'苏海恩','M',17,11);

（3）分别采用 UI 界面和 SQL 语言为 UNIVERSITY 的 Course 表输入数据。数据显示结果如图 3-3 所示。

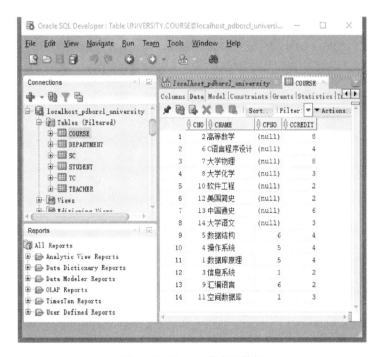

图 3-3　Course 表中的数据

插入数据的 SQL 语句如下：

insert into Course(Cno,Cname,Ccredit)values(2,'高等数学',8);
insert into Course(Cno,Cname,Ccredit)values(6,'C 语言程序设计',4);
insert into Course(Cno,Cname,Ccredit)values(7,'大学物理',8);
insert into Course(Cno,Cname,Ccredit)values(8,'大学化学',3);
insert into Course(Cno,Cname,Ccredit)values(10,'软件工程',2);
insert into Course(Cno,Cname,Ccredit)values(12,'美国简史',2);
insert into Course(Cno,Cname,Ccredit)values(13,'中国通史',6);

insert into Course(Cno,Cname,Ccredit)values(14,'大学语文',3);
insert into Course(Cno,Cname,Cpno,Ccredit)values(5,'数据结构',6,4);
insert into Course(Cno,Cname,Cpno,Ccredit)values(4,'操作系统',5,4);
insert into Course(Cno,Cname,Cpno,Ccredit)values(1,'数据库原理',5,4);
insert into Course(Cno,Cname,Cpno,Ccredit)values(3,'信息系统',1,2);
insert into Course(Cno,Cname,Cpno,Ccredit)values(9,'汇编语言',6,2);
insert into Course(Cno,Cname,Cpno,Ccredit)values(11,'空间数据库',1,3);

(4)分别采用 UI 界面和 SQL 语言为 UNIVERSITY 的 Teacher 表输入数据。数据显示结果如图3-4所示。

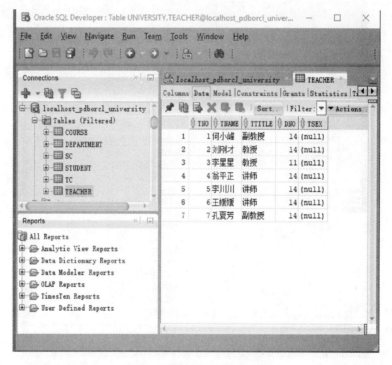

图 3-4 Teacher 表中的数据

插入数据的 SQL 语句如下：

insert into Teacher(Tno,Tname,Ttitle,Dno)values(1,'何小峰','副教授',14);
insert into Teacher(Tno,Tname,Ttitle,Dno)values(2,'刘刚才','教授',14);
insert into Teacher(Tno,Tname,Ttitle,Dno)values(3,'李星星','教授',11);
insert into Teacher(Tno,Tname,Ttitle,Dno)values(4,'翁平正','讲师',14);
insert into Teacher(Tno,Tname,Ttitle,Dno)values(5,'李川川','讲师',14);
insert into Teacher(Tno,Tname,Ttitle,Dno)values(6,'王嫒嫒','讲师',14);
insert into Teacher(Tno,Tname,Ttitle,Dno)values(7,'孔夏芳','副教授',14);

(5)分别采用 UI 界面和 SQL 语言为 UNIVERSITY 的 SC 表输入数据。数据显示结果如图 3-5 所示。

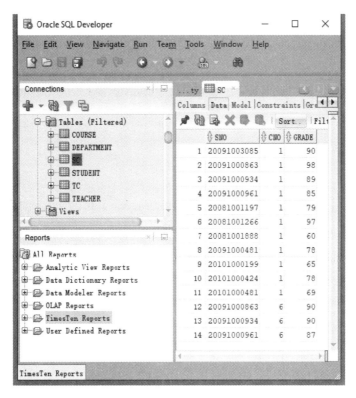

图 3-5　SC 表中的数据

插入数据的 SQL 语句如下:

insert into SC values(20091003085,1,90);

insert into SC values(20091000863,1,98);

insert into SC values(20091000934,1,89);

insert into SC values(20091000961,1,85);

insert into SC values(20081001197,1,79);

insert into SC values(20081001266,1,97);

insert into SC values(20081001888,1,60);

insert into SC values(20091000481,1,78);

insert into SC values(20101000199,1,65);

insert into SC values(20101000424,1,78);

insert into SC values(20101000481,1,69);

insert into SC values(20091000863,6,90);

insert into SC values(20091000934,6,90);

insert into SC values(20091000961,6,87);

(6) 分别采用 UI 界面和 SQL 语言为 UNIVERSITY 的 TC 表输入数据。数据显示结果如图 3-6 所示。

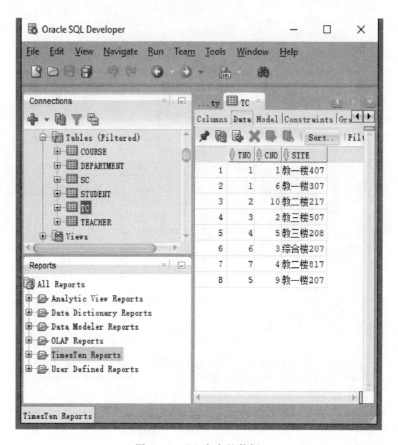

图 3-6 TC 表中的数据

插入数据的 SQL 语句如下：

insert into TC values(1,1,'教一楼 407');
insert into TC values(1,6,'教一楼 307');
insert into TC values(2,10,'教二楼 217');
insert into TC values(3,2,'教三楼 507');
insert into TC values(4,5,'教三楼 208');
insert into TC values(6,3,'综合楼 207');
insert into TC values(7,4,'教二楼 817');
insert into TC values(5,9,'教一楼 207');

(7) 采用 SQL 语言查询所有的学生信息。

select * from student;

(8)采用 SQL 语言查询所有女生的姓名。

select * from student where ssex='F';

(9)采用 SQL 语言查询各个院系学生人数。

select dno,count(*)from student group by dno;

(10)采用 SQL 语言查询各个院系老师人数。

select dno,count(*)from Teacher group by dno;

(11)采用 SQL 语言查询所有选修了"数据库原理"并且成绩在 60~100 分之间的学生的姓名和成绩,并按照成绩的降序排列。

select student. sname,sc. grade
　　from student,sc,course
　　　　where student. sno＝sc. sno
　　　　　and sc. cno＝course. cno
　　　　　and course. cname＝'数据库原理'
　　　　　and sc. grade＞60 and sc. grade＜100
　　　　　order by sc. grade desc;

(12)采用 SQL 语言编写一个连接查询:查询经济管理学院年龄在 20 岁以下的男生的姓名和年龄。

select student. sname,student. sage
　　from student,department
　　　　where student. dno＝department. dno
　　　　　and department. dname＝'经济管理学院';

(13)采用 SQL 语言编写一个嵌套查询:查询选修课程总学分在 5 个学分以上的学生的姓名。

select Sname
　　from Student
　　　　where Sno in(
　　　　　select Sno
　　　　　　from SC,Course
　　　　　　　where SC. Cno＝Course. Cno
　　　　　　　　group by Sno having SUM(CCredit)＞＝5
　　　　　　);

或者
```
select student.sname
    from student
        where student.sno in(
            select sc.sno
                from sc,
                (select sc.sno t1,sum(course.ccredit)t2
                    from sc,course
                        where sc.cno=course.cno
                            group by sc.sno)t
                where sc.sno=t.t1 and t.t2>5
            );
```

或者
```
select student.sname
    from student,
        (select sc.sno t1,sum(course.ccredit)t2
            from sc,course
                where sc.cno=course.cno
                    group by sc.sno)t
        where student.sno=t.t1 and t.t2>5;
```

(14)采用SQL语言编写一个嵌套查询:查询各门课程的最高成绩的学生姓名及其成绩。

```
select Course.cname,Student.Sname,SCX.Grade
    from Student,Course,SC SCX
        where
            Student.Sno=SCX.Sno
            and
            SCX.cno=Course.cno
            and SCX.Grade in
                (
                select MAX(Grade)
                    from SC SCY
                        where SCX.Cno=SCY.Cno
                            group by Cno
                );
```

(15)采用SQL语言查询所有选修了何小峰老师开设课程的学生姓名及其所在的院系名称。

```
select Sname,Dname
    from Student,Department,SC
        where Student.Sno=SC.Sno
        and Student.Dno=Department.Dno
        and SC.Cno in(
            select Cno
                from TC
                    where TC.Tno=
                        (select Tno
                            from Teacher
                                where Tname='何小峰')
        );
```

(16)采用 SQL 语言,在数据库中删除学号为 20091003085 的学生的所有信息(包括其选课记录)。

```
delete from SC where SC.Sno='20091003085';
delete from Student where Sno='20091003085';
```

(17)采用 SQL 语言,将学号为 20091000863 的学生的"数据库原理"这门课的成绩修改为 80 分。

```
update SC set Grade=80
    where Sno='20091000863'
        and Cno=(
            select Cno
                from Course
                    where Cname='数据库原理');
```

3.4 实验报告

实验报告按照附录 1 的格式进行编写(斜体字表示学生实验报告中要编写或要填写的内容)。

4 视图的创建与使用

4.1 实验目的

通过实习了解并掌握视图的创建、删除操作,以及视图数据的插入、修改等数据操作。
(1)通过数据库管理系统软件提供的操作界面完成视图创建、删除及其相关数据操作。
(2)通过 SQL 语言完成视图创建、删除及其相关数据操作。

4.2 实验平台

(1)操作系统:Windows XP、Windows Server 2003 及后续版本、Windows 7 及后续版本。
(2)数据库管理系统:根据实际情况,自己选择 Oracle 或 SQL Server 或 MySQL 中的一种数据库管理系统软件。
(3)本实验以第 2、第 3 章构建的 UNIVERSITY 数据库为例进行实验。

4.3 实验内容

(1)采用数据库管理系统的相应的操作界面创建一个计算机学院的学生信息视图 CSS,并新加入一条学生记录:

| 20191000911 | 钟晓年 | M | 16 |

然后,修改该学生的年龄为 18 岁,接着删除该记录,最后删除该视图。
第一步,创建视图。右键单击"Views"弹出菜单,点击"New View..."弹出如图 4-1 所示的创建视图的界面。在其中输入构建视图的查询语句,点击"确定"就可以生成视图。
第二步,在 CSS 视图中插入数据,如图 4-2 所示,点击"提交"(或者按"F11")。该项输入数据在视图中会消失,但是我们可以在 Student 数据表中看到刚才输入的数据记录,如图 4-3 所示。其中该记录的 Dno 为空值,说明 Oracle 在视图插入数据的时候,只是把 Dno 设置为空,而并没有将其设置为 14。这也说明了有的数据库管理系统(如 Oracle)在处理视图的时候并没有完全按照 SQL 语言标准进行处理。如果要其在 CSS 视图中显示,需要手工在 Student 数据表中修改该记录的 Dno 值设置为 14。

4 视图的创建与使用

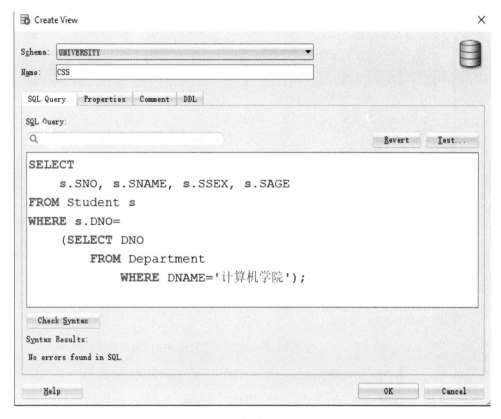

图 4-1 创建 CSS 视图

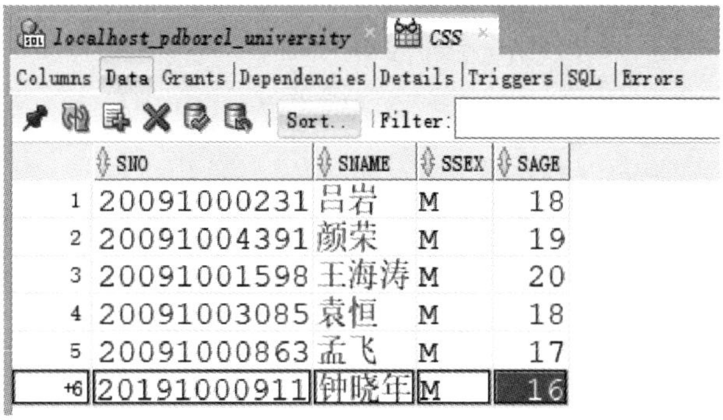

图 4-2 在视图中插入数据

第三步,在视图 CSS 的 Data 选项卡中,双击第一条记录的 SAGE 字段,输入 18,点击"F11"提交修改即可(图 4-4)。然后,可以到 Student 的数据视图中查看修改结果。

第四步,在视图 CSS 的 Data 选项卡中,选中第一条记录,点击"删除"按钮,再点击"F11"提交即可(图 4-5)。然后,到 Student 的数据视图中查看,该条记录已经不存在。

第五步,选中视图 CSS,单击右键弹出菜单,点击"删除"(Drop...)。

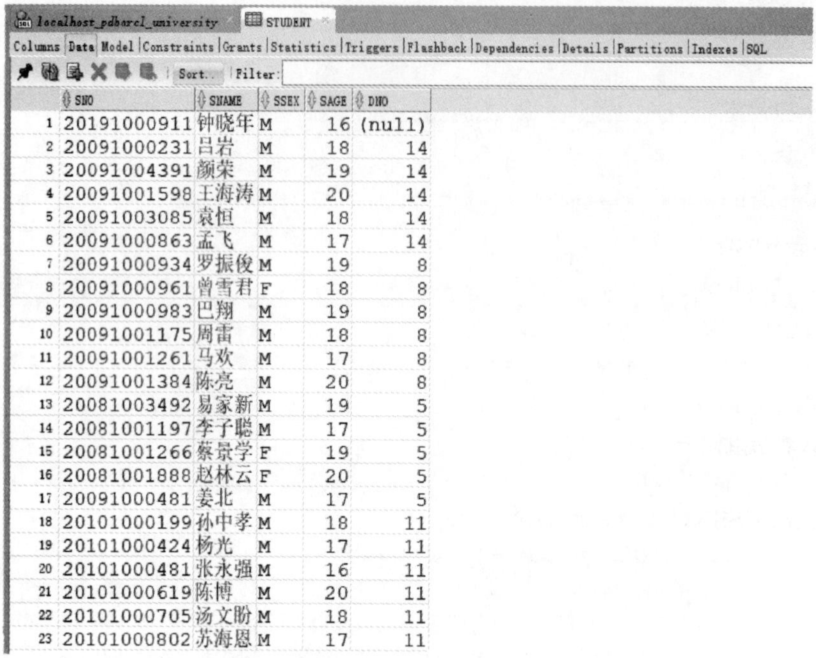

	SNO	SNAME	SSEX	SAGE	DNO
1	20191000911	钟晓年	M	16	(null)
2	20091000231	吕岩	M	18	14
3	20091004391	颜荣	M	19	14
4	20091001598	王海涛	M	20	14
5	20091003085	袁恒	M	18	14
6	20091000863	孟飞	M	17	14
7	20091000934	罗振俊	M	19	8
8	20091000961	曾雪君	F	18	8
9	20091000983	巴翔	M	19	8
10	20091001175	周雷	M	18	8
11	20091001261	马欢	M	17	8
12	20091001384	陈亮	M	20	8
13	20081003492	易家新	M	19	5
14	20081001197	李子聪	M	17	5
15	20081001266	蔡景学	F	19	5
16	20081001888	赵林云	F	20	5
17	20091000481	姜北	M	17	5
18	20101000199	孙中孝	M	18	11
19	20101000424	杨光	M	17	11
20	20101000481	张永强	M	16	11
21	20101000619	陈博	M	20	11
22	20101000705	汤文盼	M	18	11
23	20101000802	苏海恩	M	17	11

图 4－3　通过视图插入的数据记录在数据表中的显示

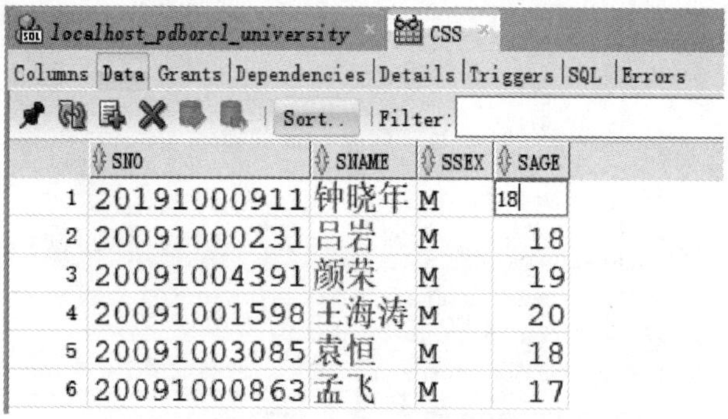

	SNO	SNAME	SSEX	SAGE
1	20191000911	钟晓年	M	18
2	20091000231	吕岩	M	18
3	20091004391	颜荣	M	19
4	20091001598	王海涛	M	20
5	20091003085	袁恒	M	18
6	20091000863	孟飞	M	17

图 4－4　在视图中修改数据

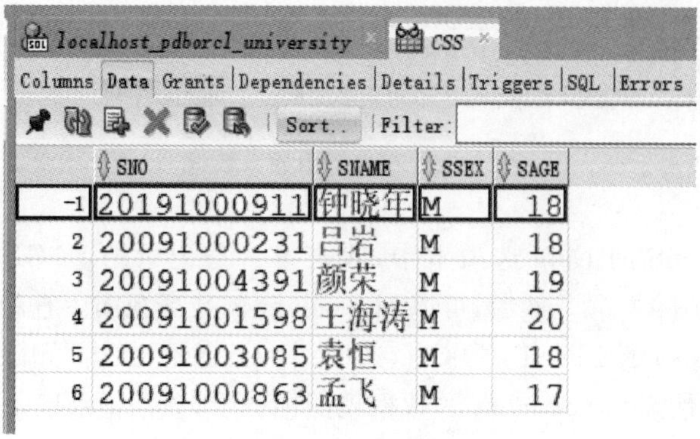

	SNO	SNAME	SSEX	SAGE
-1	20191000911	钟晓年	M	18
2	20091000231	吕岩	M	18
3	20091004391	颜荣	M	19
4	20091001598	王海涛	M	20
5	20091003085	袁恒	M	18
6	20091000863	孟飞	M	17

图 4－5　在视图中删除数据

(2)采用 SQL 语言完成上一条实验内容。

第一步,创建视图 CSS。

create view CSS(Sno,Sname,Ssex,Sage)as
  select s. Sno,s. Sname,s. Ssex,s. Sage
    from Student s
      where s. Dno=
      (Select Dno
        From Department
          where Dname='计算机学院');

第二步,在视图 CSS 中插入数据。

insert into CSS values(20191000911,'钟晓年','M',16);

第三步,在视图 CSS 中修改数据。

update student set Dno=14 where Sno=20191000911;
update CSS set Sage=18 where Sno=20191000911;

第四步,在视图 CSS 中删除数据。

delete from CSS where Sno=20191000911;

第五步,删除视图 CSS。

drop view CSS;

(3)采用 SQL 语言创建一个计算机学院教师视图 CST。

create view CST(Tno,Tname,Ttitle)as
  select t. Tno,t. Tname,t. Ttitle
    from teacher t
      where t. Dno=
      (select Dno
        from Department
          where Dname='计算机学院');

(4)采用 SQL 语言构建一个用到 CSS 和 CST 视图的查询:查询所有选修了计算机学院老师开设课程的计算机学院的学生姓名。

select Sname from CSS where Sno in
　　(select Sno from SC SCX where SCX. Cno in(
　　　　select distinct SCY. Cno
　　　　　　from SC SCY,TC
　　　　　　　　where SCY. Cno=TC. Cno and TC. Tno in
　　　　　　　　　（select　Tno from CST)));

(5)采用 SQL 语言写出在 CST 视图中删除教师编号为 1 的记录,执行并观察结果。

delete from cst where tno=1;

执行结果如下:
Error starting at line:1 in command -
delete from cst where tno=1
Error report - ORA - 02292:integrity constraint(UNIVERSITY. SYS_C0010289)violated - child record found

这说明不能在该视图中删除教师记录,它违背了完整性约束规则。

4.4　实验报告

实验报告按照附录 1 的格式进行编写(斜体字表示学生实验报告中要编写或要填写的内容)。

5 数据库安全性

5.1 实验目的

通过实习了解并掌握利用 SQL 语言对数据库进行安全性控制的操作方法。
(1)通过数据库管理系统软件提供的操作界面完成用户、角色的创建,权限修改与删除等操作。
(2)通过 SQL 语言完成用户、角色的创建、权限修改与删除等操作。
(3)视图也是数据安全性有效实现方式之一,复习第 4 章的实验,体会其在安全性方面的作用。

5.2 实验平台

(1)操作系统:Windows XP、Windows Server 2003 及后续版本、Windows 7 及后续版本。
(2)数据库管理系统:根据实际情况,自己选择 Oracle 或 SQL Server 或 MySQL 中的一种数据库管理系统软件。
(3)本实验以第 2、第 3 章构建的 UNIVERSITY 数据库为例进行实验。

5.3 实验内容

(1)新建用户 U1、U2、U3、U4、U5、U6、U7,并将连接权限授予这七个新建用户。

-- USER U1
CREATE USER U1 IDENTIFIED BY "cug"
    DEFAULT TABLESPACE "EXAMPLE"
    TEMPORARY TABLESPACE "TEMP";
GRANT "CONNECT" TO U1;

-- USER U2
CREATE USER U2 IDENTIFIED BY "cug"
    DEFAULT TABLESPACE "EXAMPLE"

        TEMPORARY TABLESPACE "TEMP";
GRANT "CONNECT" TO U2;

-- USER U3
CREATE USER U3 IDENTIFIED BY "cug"
    DEFAULT TABLESPACE "EXAMPLE"
    TEMPORARY TABLESPACE "TEMP";
GRANT "CONNECT" TO U3;

-- USER U4
CREATE USER U4 IDENTIFIED BY "cug"
    DEFAULT TABLESPACE "EXAMPLE"
    TEMPORARY TABLESPACE "TEMP";
GRANT "CONNECT" TO U4;

-- USER U5
CREATE USER U5 IDENTIFIED BY "cug"
    DEFAULT TABLESPACE "EXAMPLE"
    TEMPORARY TABLESPACE "TEMP";
GRANT "CONNECT" TO U5;

-- USER U6
CREATE USER U6 IDENTIFIED BY "cug"
    DEFAULT TABLESPACE "EXAMPLE"
    TEMPORARY TABLESPACE "TEMP";
GRANT "CONNECT" TO U6;

-- USER U7
CREATE USER U7 IDENTIFIED BY "cug"
    DEFAULT TABLESPACE "EXAMPLE"
    TEMPORARY TABLESPACE "TEMP";
GRANT "CONNECT" TO U7;

(2)将 Student 表的查询权限授予 U1。

GRANT SELECT on "UNIVERSITY"."Student" to "U1";

(3)把 Student 表和 Course 表的全部权限授予 U2 和 U3。

GRANT all on "UNIVERSITY"."Student" to "U2";
GRANT all on "UNIVERSITY"."Course" to "U3";

(4)将 SC 表的查询权限授予所有用户。

GRANT SELECT on "UNIVERSITY"."SC" to public;

(5)将 SC 表的更新和查询权限授予 U4。

GRANT SELECT,UPDATE on "UNIVERSITY"."SC" to "U4";

(6)将 SC 表的插入权限授予 U5,并允许将此权限再授予其他用户。

GRANT INSERT on "UNIVERSITY"."SC" to "U5" WITH GRANT OPTION;

(7)用 U5 连接数据库服务器,将 SC 表的插入权限授予 U6,不允许 U6 再转授权。

GRANT INSERT on "UNIVERSITY"."SC" to "U6";

(8)用 U6 连接数据库服务器,将 SC 表的插入权限授予 U7,看看授权是否成功。

GRANT INSERT on "UNIVERSITY"."SC" to "U7";

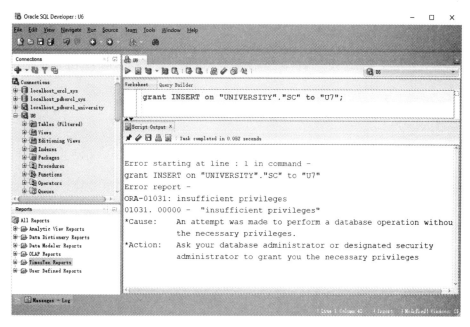

图 5-1 U6 用户没有传递授权的权限

执行结果如图 5-1 所示,系统会提示权限不够。

Error starting at line:1 in command -
grant INSERT on "UNIVERSITY"."SC" to "U7"

Error report -
ORA - 01031: insufficient privileges
01031.00000 - "insufficient privileges"
* Cause: An attempt was made to perform a database operation without
             the necessary privileges.
* Action: Ask your database administrator or designated security
             administrator to grant you the necessary privileges

(9)收回 U4 修改 SC 表的权限。

REVOKE UPDATE on SC from U4;

(10)收回所有用户对 SC 表的查询权限。

REVOKE SELECT on SC from public;

注意:如果采用的是 Oracle 12c 及其以上的版本,并且连接的不是可插拔数据库(Pluggable Database),则需要在 U1 至 U7 的所有用户前面加上 C##,也即变成 C##U1 至 C##U7。

5.4　实验报告

实验报告按照附录 1 的格式进行编写(斜体字表示学生实验报告中要编写或要填写的内容)。

6 数据库完整性

6.1 实验目的

通过实习掌握完整性的定义与维护方法,重点掌握利用 SQL 语言对数据库进行完整性控制,包括实体完整性、参照完整性和用户自定义完整性的操作。

6.2 实验平台

(1)操作系统:Windows XP、Windows Server 2003 及后续版本、Windows 7 及后续版本。
(2)数据库管理系统:根据实际情况,自己选择 Oracle 或 SQL Server 或 MySQL 中的一种数据库管理系统软件。
(3)本实验以第 2、第 3 章构建的 UNIVERSITY 数据库为例进行实验。

6.3 实验内容

(1)在 UNIVERSITY 数据库中创建表 STU_T,该表与 Student 表具有相同字段,其主码为 Sno。
(a)创建 STU_T 表时定义实体完整性(列级实体完整性),然后删除 STU_T。

```
create table   STU_T  (
   Sno int primary key(Sno)
   Sname varchar(50),
   Ssex varchar(2),
   Sage int,
   Dno int
)  ;
   drop table   stu_t;
```

(b)创建 STU_T 表时定义实体完整性(表级实体完整性),然后删除 STU_T。

```
create table   STU_T  (
```

    Sno int,
    Sname varchar(50),
    Ssex varchar(2),
    Sage int,
    Dno int,
    primary key(Sno)
) ;
drop table STU_T;

(c)创建 STU_T 表后再定义其实体完整性 PK_SNO(提示：采用 alter 命令添加实体完整性)。

create table STU_T (
    Sno int,
    Sname varchar(50),
    Ssex varchar(2),
    SagE int,
    Dno int
) ;
alter table STU_T add constraint PM_SNO primary key(Sno);

(d)删除数据表 STU_T 中的实体完整性 PK_SNO,然后删除 STU_T。

alter table STU_T drop constraint PK_SNO;
drop table STU_T;

(2)在 UNIVERSITY 数据库中创建表 SC_T,该表与 SC 表具有相同字段,其主码为 Sno、Cno。

(a)创建 SC_T 表时定义实体完整性(表级实体完整性),然后删除 SC_T。

create table SC_T(
    Sno int,
    Cno int,
    Grade float,
    primary key(Sno,Cno));
drop table SC_T;

(b)创建 SC_T 表后再定义其实体完整性 PK_SC(提示：采用 alter 命令添加实体完整性),然后删除数据表 SC_T 中的实体完整性 PK_SC,并删除 SC_T。

create table SC_T(

        Sno int,
        Cno int,
        Grade float);
alter table SC_T add constraint PK_SC primary key(Sno,Cno);
alter table SC_T drop constraint PK_SC;
drop table SC_T;

(3)在 UNIVERSITY 数据库中创建表 TC_T,该表与 TC 表具有相同字段,其主码为 Cno、Tno。

(a)创建 TC_T 表时定义实体完整性和参照完整性(采用表级实体完整性和参照完整性,被参照的数据表分别为 Teacher 表中的 Tno 和 Course 表中的 Cno),然后删除 TC_T 表。

    create table TC_T(Tno INT,Cno INT,Site VARCHAR(50),
        primary key(Tno,Cno),
        foreign key(Tno)references   Teacher(Tno),
        foreign key(Cno)references   Course(Cno));
    drop table TC_T;

(b)创建 TC_T 表后再定义其实体完整性和参照完整性(提示:采用 alter 命令增加实体完整性和参照完整性,被参照的数据表分别为 Teacher 表中的 Tno 和 Course 表中的 Cno)。

    create table TC_T(Tno int,Cno int,Site varchar(50));
    alter table TC_T add constraint PK_TC primary key(Tno,Cno);
    alter table TC_T add constraint FK_Tno foreign key(Tno)references   Teacher(Tno);
    alter table TC_T add constraint FK_Cno foreign key(Cno)references   Course(Cno);
    drop table TC_T;

(4)在 UNIVERSITY 数据库中创建表 DEP_T,该表与 Department 表具有相同字段,其主码为 Dno。

(a)创建 DEP_T 表时定义实体完整性(列级实体完整性),并定义 Dname 唯一且非空,Daddress 的默认值为"北一楼",然后删除 DEP_T。

    create table DEP_T(
        Dno int primary key,
        Dname varchar(50)unique not null,
        Daddress varchar(50)default '北一楼');
    drop table   DEP_T cascade constraints purge;

(b)创建 DEP_T 表时定义实体完整性(表级实体完整性),并限制 Dno 的取值范围为 00~99(提示:使用 Check 进行约束),然后删除 DEP_T。

```
create table DEP_T(
    Dno int,
    Dname varchar(50),
    Daddress varchar(50),
    primary key(Dno),
    check(Dno>0 and Dno<100));
drop table   DEP_T cascade constraints purge;
```

6.4 实验报告

实验报告按照附录1的格式进行编写(斜体字表示学生实验报告中要编写或要填写的内容)。

7 触发器

触发器是用户定义在表上的一类由事件驱动的特殊过程。触发器一旦被定义,它将被保存在数据库服务器中。任何用户对表的增加、删除、修改操作均由服务器自动激活相应的触发器。

7.1 实验目的

掌握数据库触发器的设计与使用方法。

7.2 实验平台

(1)操作系统:Windows XP、Windows Server 2003 及后续版本、Windows 7 及后续版本。

(2)数据库管理系统:根据实际情况,自己选择 Oracle 或 SQL Server 或 MySQL 中的一种数据库管理系统软件(注意:不同的数据库管理系统的触发器有较大差别,其通用性并不太好)。

(3)本实验以第 2、第 3 章构建的 UNIVERSITY 数据库为例进行实验。

7.3 实验内容

(1)在 UNIVERSITY 数据库中创建表 SGA_T,记录每个学生的所有选修课程的平均成绩,其主要字段包括 Sno、AverageGrade,其主码为 Sno。SGA_T 表的创建与初始化 SQL 语句如下(图 7 - 1):

create table SGA_T(Sno int primary key,AverageGrade float);
insert into SGA_T select Sno,avg(Grade)from SC group by Sno;

(a)在 SC 表上定义一个 update 触发器,当修改某个学生的某一门选修课程的成绩后,自动重新计算所有的平均成绩,并更新到 SGA_T 表中。

第一步,对于触发器的创建可以采用界面进行操作,也可编写 SQL 语句进行操作。两者在本质上其实是一样的。这里采用界面操作,如图 7 - 2 所示。在 SC 表节点上点击右键弹出菜单,点击"Trigger"中的"Create...",弹出如图 7 - 3 所示的触发器创建对话框。

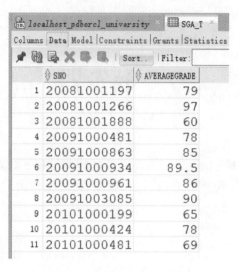

图 7-1 初始化的 SGA_T 表

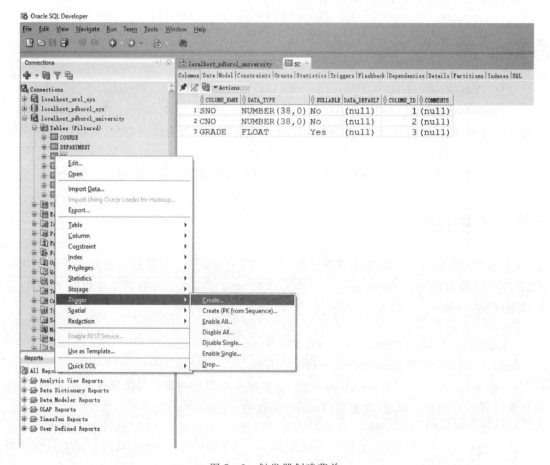

图 7-2 触发器创建菜单

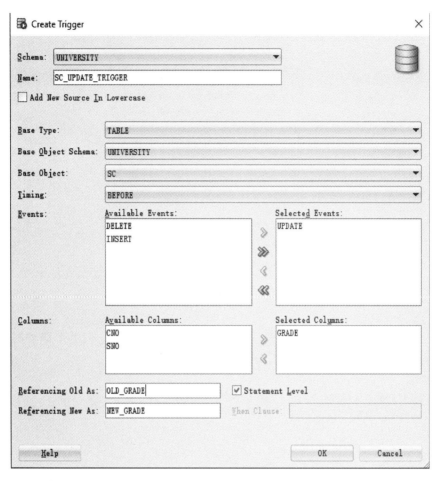

图 7-3 创建触发器对话框

第二步，填写如图 7-3 所示的相关参数后，生成如图 7-4 中所示的初始化代码。

create or replace trigger SC_UPDATE_TRIGGER
before update of Grade on SC
referencing old as old_grade new as new_grade
begin
  null;
end;

第三步，在 begin 和 end 之间编写处理代码，实现修改后自动更新 SGA_T 中的平均成绩数据。具体实现代码如下：

delete from SGA_T;
insert into SGA_T select Sno,avg(Grade)from SC group by Sno;

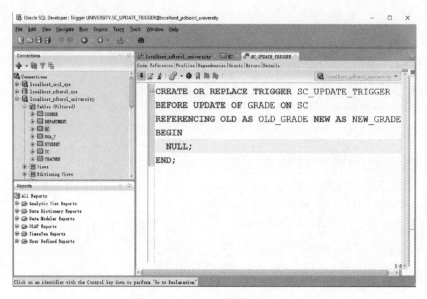

图 7-4 生成的触发器初始代码

(b)在 SC 表上定义一个 insert 触发器,当添加某个学生的某一门选修课程的成绩时,自动重新计算所有学生的平均成绩,并更新到 SGA_T 表中。

create or replace trigger SC_INSERT_TRIGGER
after insert on SC
begin
　delete from SGA_T;
　insert into SGA_T select Sno,avg(Grade)from SC group by Sno;
end;

(c)在 SC 表上定义一个 delete 触发器,当删除某个学生的某一门选修课程的成绩时,自动重新计算所有学生平均成绩,并更新到 SGA_T 表中。

create or replace trigger SC_DELETE_TRIGGER
after delete on SC
begin
　delete from SGA_T;
　insert into SGA_T select Sno,avg(Grade)from SC group by Sno;
end;

(d)删除上面三个触发器,并删除数据表 SGA_T。

drop trigger SC_INSERT_TRIGGER;
drop trigger SC_UPDATE_TRIGGER;

drop trigger SC_DELETE_TRIGGER;
drop table SGA_T;

(2)在 UNIVERSITY 数据库中,我们规定每个老师至少必须上一门课程,最多只能同时上三门课程。

(a)在 TC 表上定义一个 INSERT 触发器,当添加某个老师新上的一门课程时,先检查该教师所上的课程总数是否达到了三门,如果达到三门的上限,则提示不能再为该教师新增课程。

```
create or replace trigger TC_INSERT_TRIGGER
before insert on TC
for each row
declare
    cc int;
begin
    select count(*)into cc from TC where TC.Tno=:NEW.Tno;
    if cc>=3 then
        dbms_output.put_line('NO ACTION');
    end if;
end;
```

(b)在 TC 表上定义一个 delete 触发器,当删除某个老师所上的某一门课程时,先检查该教师所上的课程总数是否只有一门,如果是,则提示不能再为该教师减少课程。

```
create or replace trigger TC_DELETE_TRIGGER
before delete on TC
for each row
declare
    cc int;
begin
    select count(*)into cc from TC where TC.Tno=:NEW.Tno;
    if cc<=1 then
        dbms_output.put_line('NO ACTION');
    end if;
end;
```

(c)删除上面定义的两个触发器。

drop trigger TC_INSERT_TRIGGER;
drop trigger TC_DELETE_TRIGGER;

(3)在 UNIVERSITY 数据库中,我们规定每个老师只能属于一个系,在一个老师从一个系调动到另外一个系之前,需要先检查接受调动的系的编号是否合法。请在 Teacher 表上设计一个 before update 触发器来实现上述功能,然后删除该触发器。

```
create or replace trigger TEACHER_TRIGGER
before update of dno on Teacher
for each row
declare
    cc int;
begin
   select count(*)into cc from Department where Dno=:NEW.Dno;
   if cc<=0   then
     dbms_output.put_line('NO ACTION');
   end if;
end;

drop trigger TEACHER_TRIGGER;
```

7.4 实验报告

实验报告按照附录 1 的格式进行编写(斜体字表示学生实验报告中要编写或要填写的内容)。

8 数据库设计

数据库设计与开发时常是联系在一起的,也是一个很大的项目。本实验重点讨论的是数据库的概念设计、逻辑设计和物理设计。

8.1 实验目的

掌握数据库设计的基本方法及数据库设计工具。具体而言,需要掌握数据库设计基本步骤,包括数据库概念结构设计、数据库逻辑结构设计、数据库物理结构设计以及相关 SQL 语句生成。至少掌握一种 E-R 绘制工具,能将设计的 E-R 图绘制、表达出来。

8.2 实验平台

(1)操作系统:Windows XP、Windows Server 2003 及后续版本、Windows 7 及后续版本。
(2)数据库管理系统:根据实际情况,自己选择 Oracle 或 SQL Server 或 MySQL 中的一种数据库管理系统软件。
(3)本实验以第 2、第 3 章构建的 UNIVERSITY 数据库为例进行实验。
(4)E-R 图绘制工具,如 Microsoft Office Visio、PowerDesigner、ERWin、Oracle Data Modeler、Microsft Visual Studio 等具备 E-R 绘制功能的软件。

8.3 实验内容

设计一个大学教学信息管理应用数据库 UNIVERSITY。其中,一个教师属于一个系,一个系有多名教师,每个系都有自己的办公地点。每个教师可以讲授多门课程,每个学生属于一个系,可以选修多门课程。每门课程具有一定学分,并可能有先导课程。

1. 数据库概念结构设计

识别出教师(Teacher)、系(Department)、课程(Course)、学生(Student)四个实体。每个实体的属性和码如下。

(1)系(Department)。系的编号为 Dno,系的名称为 Dname,系所在的办公地址为 Daddress。主码为系的编号 Dno。
(2)学生(Student)。学生学号为 Sno,学生姓名为 Sname,学生性别为 Ssex,学生年龄为

Sage,学生所属系编号 Dno。主码为学生学号 Sno。

(3)教师(Teacher)。教师编号为 Tno,教师姓名为 Tname,教师职称为 Ttitle。主码为教师编号 Tno。

(4)课程(Course)。课程编号为 Cno,课程名称为 Cname,先导课程编号为 Cpno,课程学分为 Ccredit。主码为课程编号 Cno。

根据实际语义,分析实体之间的联系,确定实体之间一对一、一对多和多对多联系,绘制 E-R 图(图 8-1)。

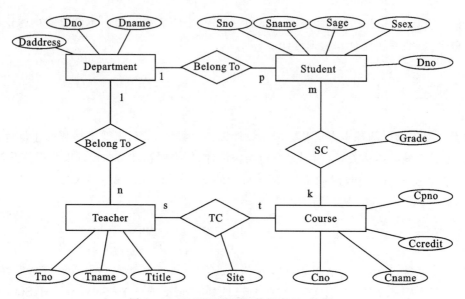

图 8-1　UNIVERSITY 数据库 E-R 图

2. 数据库逻辑结构设计

按照数据库设计中概念结构向逻辑结构转换规则,根据所绘制 E-R 图,设计 UNIVERSITY 数据库逻辑结构(列出所有关系,确定每个字段的类型、长度等信息,以表的形式列出,画出数据库模式图),并写出相关 SQL 语句。

第一步,写出 UNIVERSITY 关系数据库模式。

(1)系的信息表 Department(<u>Dno</u>,Dname,Daddress)。

(2)学生信息表 Student(<u>Sno</u>,Sname,Ssex,Sage,Dno)。

(3)教师信息表 Teacher(<u>Tno</u>,Tname,Ttitle,Dno)。

(4)课程信息表 Course(<u>Cno</u>,Cname,Cpno,Ccredit)。

(5)学生选课表 SC(<u>Sno</u>,<u>Cno</u>,Grade)。

(6)教师授课表 TC(<u>Tno</u>,<u>Cno</u>,Site)。

第二步,结合选定的数据库管理系统(Oracle),列出每个关系中每个属性的类型、长度等信息(表 8-1~表 8-5)。

表 8-1 Department 关系属性表

关系名称		Department		关系别名		系的信息	
属性名	别名	类型	长度	值域	唯一	可空	备注
Dno	系编号	INT			Y	N	
Dname	系名称	VARCHAR	50		Y	N	
Daddress	系地址	VARCHAR	50		Y	Y	

表 8-2 Student 关系属性表

关系名称		Student		关系别名		学生信息	
属性名	别名	类型	长度	值域	唯一	可空	备注
Sno	学号	INT			Y	N	
Sname	姓名	VARCHAR	50		N	N	
Ssex	性别	VARCHAR	2	M/F	N	Y	
Sage	年龄	INT			N	Y	
Dno	系编号	INT			Y	Y	所在系

表 8-3 Teacher 关系属性表

关系名称		Teacher		关系别名		教师信息	
属性名	别名	类型	长度	值域	唯一	可空	备注
Tno	工号	INT			Y	N	
Tname	姓名	VARCHAR	50		N	N	
Ttitle	性别	VARCHAR	50		N	Y	
Dno	系编号	INT			Y	Y	所在系

表 8-4 Course 关系属性表

关系名称		Course		关系别名		课程信息	
属性名	别名	类型	长度	值域	唯一	可空	备注
Cno	课程号	INT			Y	N	
Cname	课程名	VARCHAR	50		Y	N	
Cpno	先导课	INT			N	Y	
Ccredit	学分	INT			Y	Y	所在系

表 8-5 SC 关系属性表

关系名称		SC		关系别名		学生选课信息	
属性名	别名	类型	长度	值域	唯一	可空	备注
Cno	课程号	INT			Y	N	
Sno	学号	INT			Y	N	
Grade	成绩	FLOAT			N	Y	

表 8-6 TC 关系属性表

关系名称		TC		关系别名		教师授课信息	
属性名	别名	类型	长度	值域	唯一	可空	备注
Cno	课程号	INT			Y	N	
Tno	工号	INT			Y	N	
Site	位置	VARCHAR	50		N	Y	

第三步，画出数据库模式图，如图 8-2 所示（采用 Oracle Data Modeler 绘制）。需要说明的是，在 Oracle 中，INT 类型会被转换成 NUMBER(38)，VARCHAR 类型会被转换成 VARCHAR2，FLOAT 也会转换成 NUMBER 类型。

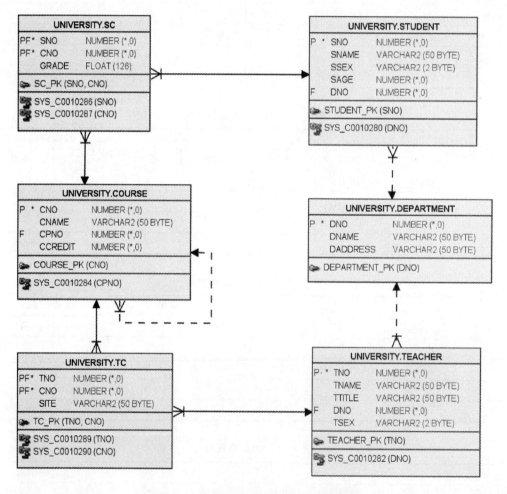

图 8-2 UNIVERSITY 关系数据模型

第四步，设计用户子模式。例如，设计计算机系学生信息视图 CSS。这里没有需要设计的其他用户子模式，故可以直接进入下一步物理设计。

3. 数据库物理结构设计

数据库物理结构设计首先根据数据库逻辑结构自动转换生成,然后再根据实际应用需求,设计数据库的索引与存储结构。这里主要给出数据库结构的 SQL 语言实现。

```
create table Department(
    Dno int,
    Dname varchar(50),
    Daddress varchar(50),
    primary key(Dno)
);
create table Student(
    Sno int,
    Sname varchar(50),
    Ssex varchar(2),
    Sage int,
    Dno int,
    primary key(Sno),
    foreign key(Dno)references Department(Dno)
);
create table Teacher(
    Tno int primary key,
    Tname varchar(50),
    Ttitle varchar(50),
    Dno int,
    foreign key(Dno)references Department(Dno)
);
create table Course(
    Cno int primary key,
    Cname varchar(50),
    Cpno int,
    CCredit int,
    foreign key(Cpno)references Course(Cno)
);
create table SC(
    Sno int,
    Cno int,
    Grade float,
    primary key(Sno,Cno),
```

```
        foreign key(Sno)references Student(Sno),
        foreign key(Cno)references Course(Cno)
);
create table TC(
        Tno int,
        Cno int,
        Site varchar(50),
        primary key(Tno,Cno),
        foreign key(Tno)references Teacher(Tno),
        foreign key(Cno)references Course(Cno)
);
```

8.4 实验报告

实验报告按照附录 1 的格式进行编写(斜体字表示学生实验报告中要编写或要填写的内容)。

9 存储过程与函数

本实验包括存储过程、函数与游标三部分内容,是数据库服务器端的程序开发的基础。所谓存储过程是由过程化 SQL 语句编写的过程,该过程经过编译和优化处理后存储在数据库服务器中,在使用时可以随时调用。函数与存储过程类似,也是用户定义的持久化存储模块,不同之处在于函数需要指定返回类型。游标是系统为用户开设的一个数据缓冲区,存放 SQL 语句的执行结果。

9.1 实验目的

(1)掌握数据库 PL/SQL 编程语言基础。
(2)掌握数据库存储过程的设计与使用方法,主要包括存储过程定义、运行、变更、删除以及参数传递等。
(3)掌握数据库自定义函数的设计与使用方法,主要包括函数的定义、运行、变更、删除以及参数传递等。
(4)掌握 PL/SQL 中游标的设计与使用方法。

9.2 实验平台

(1)操作系统:Windows XP、Windows Server 2003 及后续版本、Windows 7 及后续版本。
(2)数据库管理系统:根据实际情况,自己选择 Oracle 或 SQL Server 或 MySQL 中的一种数据库管理系统软件。
(3)本实验以第 2、第 3 章构建的 UNIVERSITY 数据库为例进行实验。

9.3 实验内容

(1)定义一个无参数的存储过程 DecreaseGrade,更新所有学生成绩,将其降低 5%,并调用该存储过程。

```
create or replace procedure DecreaseGrade as
begin
    update SC set Grade=Grade * 0.95;
```

```
end DecreaseGrade;
/
begin
    DecreaseGrade();
end;
```

(2)定义一个带输入参数的存储过程 IncreaseGrade,将课程号为 1 的所有学生成绩提升 5%;要求课程号作为存储过程参数传入,并调用该存储过程。

```
create or replace procedure IncreaseGrade(
    ccno in int
)as
begin
    update SC set Grade=Grade * 1.05 where Cno=ccno;
end IncreaseGrade;
/
declare
    ccno number;
begin
    ccno:=1;
    IncreaseGrade(
        ccno=> ccno
    );
end;
```

(3)定义一个带有输入和输出参数的存储过程 AverageStudentGrade,计算一个学生的所有选修课程的平均成绩,要求以学号为输入参数,以计算结果(该生的所有选修课平均成绩)为输出参数;调用该存储过程,并输出计算结果。

```
create or replace procedure AverageStudentGrade
    (
        paramsno in int,
        paramgrade out float
    )as
        g float;
    begin
        select sg.ag into g
            from(select Sno s,avg(Grade)ag
                from SC group by Sno)sg
                    where sg.s=paramsno;
```

```
        paramgrade:=g;
    end AverageStudentGrade;
/
    declare
        paramsno number;
        paramgrade float;
    begin
        paramsno:=20091000863;
        AverageStudentGrade(
            paramsno=> paramsno,
            paramgrade=> paramgrade
        );
        dbms_output.put_line('paramgrade='|| paramgrade);
    end;
```

(4)删除存储过程 IncreaseGrade、DecreaseGrades。

```
drop procedure DecreaseGrade;
drop procedure IncreaseGrade;
```

(5)定义一个带有输入参数的自定义函数 CalculateAverageStudentGrade,计算一个学生的所有选修课程的平均成绩,要求以学号为输入参数,返回该生的所有选修课平均成绩;调用函数,并输出计算结果。

```
create or replace function CalculateAverageStudentGrade
(
    paramsno in number
)return number as
g number;
begin
    select sg.ag into g
        from(select Sno s,avg(Grade)ag
                from SC group by Sno)sg
                    where sg.s=paramsno;
    return g;
end CalculateAverageStudentGrade;
/
declare
    paramsno number;
    v_return number;
```

```
begin
    paramsno:=20091000863;
    v_Return:=CalculateAverageStudentGrade(
        paramsno=> paramsno
    );
    dbms_output.put_line('v_return=' || v_return);
end;
```

(6) 删除函数 CalculateAverageStudentGrade。

```
drop function CalculateAverageStudentGrade;
```

(7) 定义一个存储过程,采用普通无参游标实现计算学校开设的所有课程的学分之和。

```
create or replace procedure pro_7(
    vresult out number
)as
    cc number;
    ss number;
    cursor cur is select Ccredit from Course;
begin
    ss:=0;
    open cur;
    loop
        fetch cur into cc;
        if cur%FOUND then
            dbms_output.put_line(cc);
            ss:=ss+cc;
        else
            dbms_output.put_line(ss);
            vresult:=ss;
            exit;
        end if;
    end loop;
end pro_7;
/
declare
    vresult number;
begin
    pro_7(
```

```
      vresult=> vresult
    );
    dbms_output.put_line('vresult=' || vresult);
end;
```

(8) 定义一个存储过程,采用 REF CURSOR 实现计算学校所有学生选修课程的成绩之和。

```
create or replace procedure pro_8
(
   vresult out number
)as
type rc is ref cursor;--动态游标
cur rc;
cc number;
ss number;
begin
   ss:=0;
   open cur for select Grade from SC;
    loop
      fetch cur into cc;
      if cur%FOUND then
         dbms_output.put_line(cc);
         ss:=ss+cc;
      else
         dbms_output.put_line(ss);
         vresult:=ss;
         exit;
      end if;
    end loop;
end pro_8;
/
declare
   vresult number;
begin
   pro_8(
      vresult=> vresult
   );
dbms_output.put_line('vresult=' || vresult);
end;
```

(9) 定义一个存储过程，采用带参数游标实现按照学号计算学生的平均成绩。

```sql
create or replace procedure pro_9
(
    sno_in in number
)as
g number;
cursor cur(paramsno number)is select sg. ag
        from(select Sno s,avg(Grade)ag
            from SC group by Sno)sg
                where sg. s=paramsno;
begin
    open cur(sno_in);
    loop
        fetch cur into g;
        if cur%FOUND then
            dbms_output. put_line(g);
        else
            dbms_output. put_line(g);
            exit;
        end if;
    end loop;
end pro_9;
/
declare
    sno_in number;
begin
    sno_in:=20091000863;
    pro_9(
        sno_in=> sno_in
    );
end;
```

9.4 实验报告

实验报告按照附录1的格式进行编写（斜体字表示学生实验报告中要编写或要填写的内容）。

10 数据库应用开发(C++)

数据库应用程序都会采用一定的方式连接数据库,从中进行数据读取与写入,或者调用数据库服务器中的存储过程或函数进行某些计算处理。采用高级语言 C/C++进行数据库应用程序开发,一般有两种连接数据库的方式。

第一种通过 ODBC(Open Database Connectivity,ODBC)进行连接。ODBC 是微软公司开放服务体系(Windows Open Services Architecture,WOSA)中有关数据库的一个组成部分,它提供了一组基于 C 语言访问数据库的应用程序编程接口(Application Programming Interface,API)。ODBC 在一定程度上规范应用开发和关系数据库管理系统应用接口,其体系结构如图 10-1 所示。

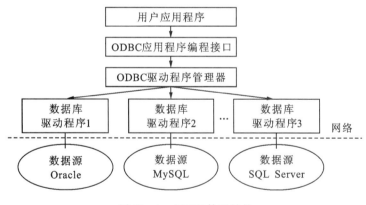

图 10-1 ODBC 体系结构

第二种是通过数据库管理系统提供的特殊连接库或组件进行数据库连接。例如,在 Oracle 中,如果采用 C/C++进行数据库应用程序开发,一般追求的都是高效率。Oracle 专门提供了功能强大的 OCI(Oracle Call Interface)和 OCCI(Oracle C++ Call Interface)来连接 Oracle 数据库。其中,OCI 是一种基于本地 C 语言的复杂、高效的 Oracle 数据库接口。Oracle 的一些工具,如 SQL * Plus、Real Application Testing(RAT)、SQL * Loader、Data - Pump 等都是采用的 OCI。OCI 也是 Oracle 数据库一些其他语言的接口的基础,例如 JDBC - OCI、Oracle Data Provider for Net(ODP. Net)、Oracle Precompilers、Oracle ODBC、PHP OCI8、ruby - oci8、Perl DBD::Oracle、Python cx_Oracle 以及 R 语言的 ROracle 等都是基于 OCI 构建的。OCCI 是对 OCI 的面向对象封装,这两种连接程序是 Oracle 数据库中提供的功能最全面、效率最高的数据库连接方式。本实验将基于 Oracle ODBC 和 OCI 分别建立数据库应用程序。

10.1 实验目的

(1)掌握基于 ODBC 的编程原理与基本步骤。
(2)掌握基于 Oracle ODBC 和 OCI 的数据库连接程序的数据库应用程序开发。

10.2 实验平台

(1)操作系统:Windows XP、Windows Server 2003 及后续版本、Windows 7 及后续版本。
(2)数据库管理系统:Oracle 11g 及后续版本。
(3)C++开发环境:Visual C++6.0 或 Visual Studio 2010 及后续版本,或 CLion 等。
(4)本实验以第 2、第 3 章构建的 UNIVERSITY 数据库为例进行实验。

10.3 实验内容

(1)配置 Oracle 数据库的 ODBC 连接,采用 C++开发一个基于 MFC 和 ODBC 的数据库应用程序,实现以下功能:输出所有学生的选修课程名称和成绩。

(a)配置 University 数据源。

第一步,启动 Oracle 数据库服务和监听服务,如果是采用的 Oracle 12c 中的可插拔数据库,请记得开启 PDB;打开 Microsoft ODBC Administrator,如图 10-2 所示。

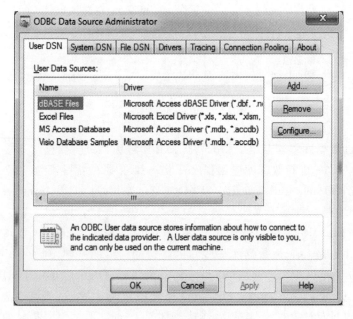

图 10-2 ODBC 数据源管理器

第二步,点击"System DSN"中的"Add…"按钮,弹出如图 10-3 所示界面,显示本机中已经存在 Oracle12c 和 SQL Server 的驱动程序。我们点击"Oracle in OraDB12Home1",单击"Finish"按钮,弹出如图 10-3 所示的 Oracle 数据源配置界面。

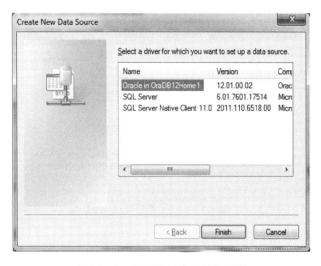

图 10-3 ODBC 创建新的数据源

第三步,配置 Oracle ODBC 驱动与数据源,如图 10-4 所示,在 Data Source Name 中填入数据源名称,自己设置一个便于自己理解的字符串就可以了,如本例中填入 UNIVERSITY_PDBORCL。在 Description 一栏填入数据源的描述信息,用于更加确切地说明 ODBC 数据源的用途等信息。TNS Service Name 必须是 Oracle 数据库服务名,由于我们连接的是可插拔数据库,这里填入 PDBORCL。再下一行为 User ID,填写连接用户名 UNIVERSITY。然后,点击"Test Connection"按钮,会提示输入密码。输入正确用户密码后,提示连接测试成功。我们就在系统中设置了一个 Oracle ODBC 的数据源 UNIVERSITY_PDBORCL,如图 10-5 所示。

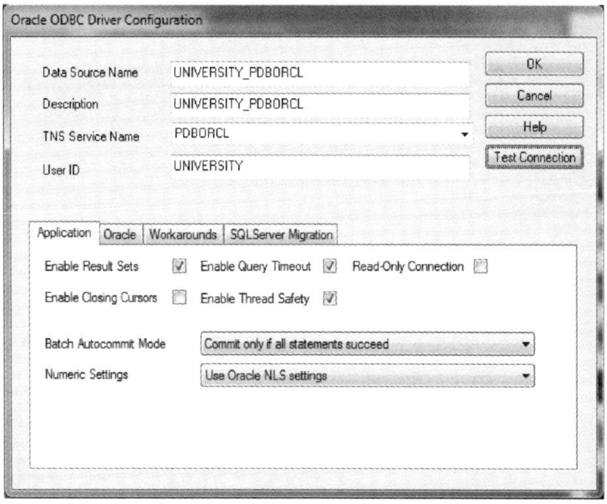

图 10-4 添加 ODBC 创建新的数据源

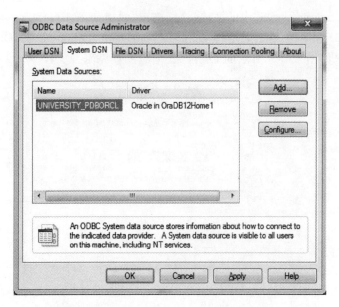

图 10-5　成功添加一个 System DSN 数据源

(b)基于 MFC ODBC 开发数据库应用程序。

由于直接采用 C 语言版的 ODBC API 编程比较繁琐,所以较多的数据库组件或库都对 ODBC 的 C 语言版本 API 进行了面向对象的封装。其中,MFC 中的 CDatabase、CRecordset 等类封装 C 语言版的 ODBC API 的大部分功能。为简便起见,我们这里直接采用 Visual Studio 2017 社区版进行基于 ODBC 的数据库应用程序开发。

第一步,按照图 10-6～图 10-11 的方式生成示例工程。

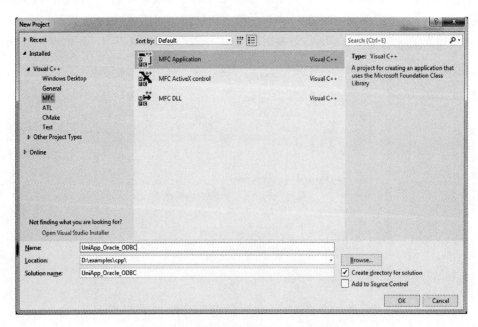

图 10-6　选择 MFC Application 工程类型

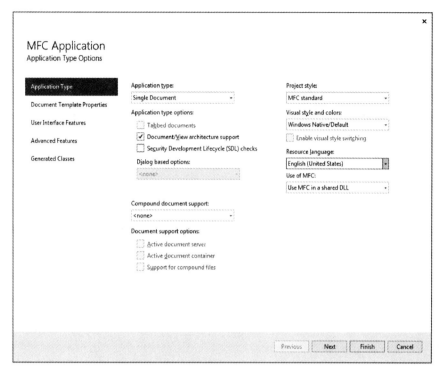

图 10-7　选择 Single Document 应用程序类型

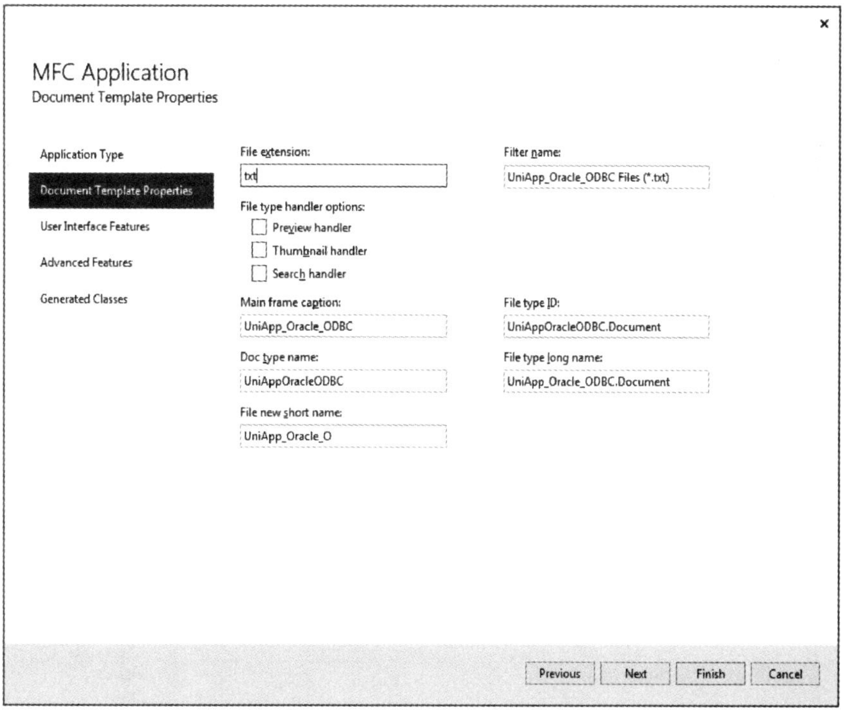

图 10-8　选择默认的文档模板属性

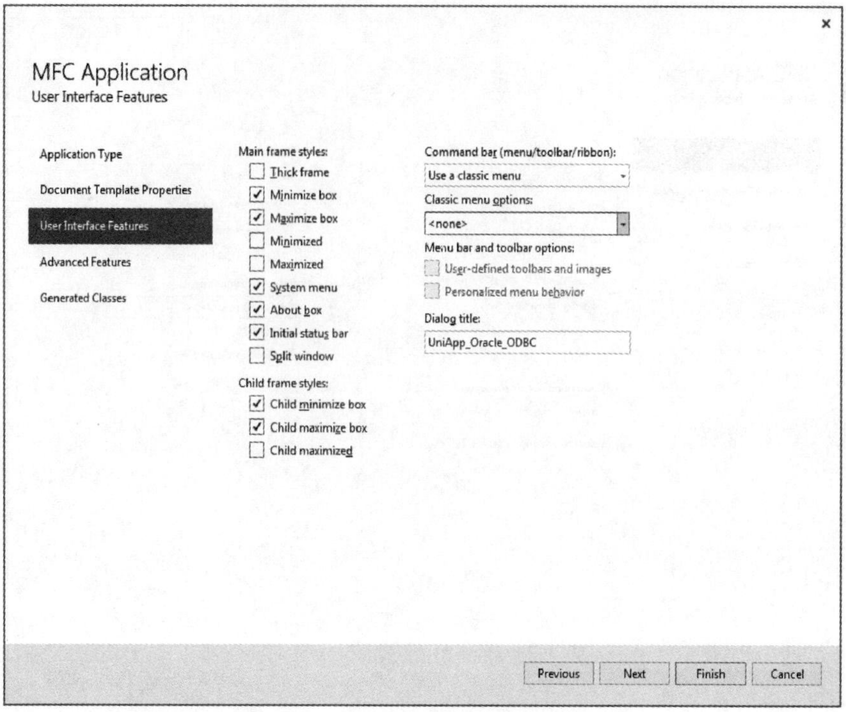

图 10－9　选择用户界面特征

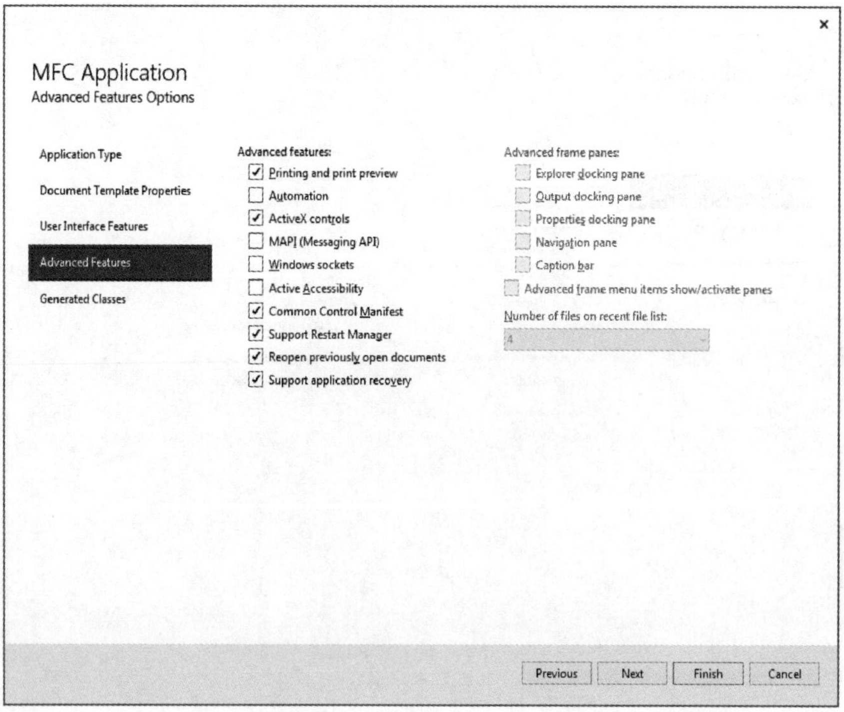

图 10－10　选择高级特征

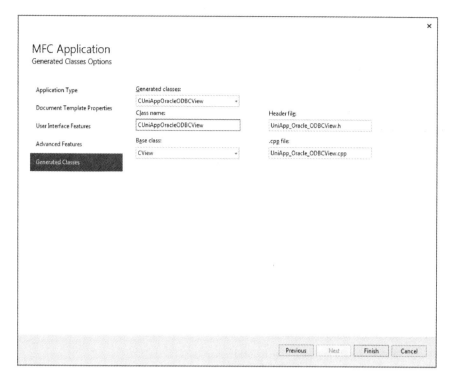

图 10-11 采用默认的生成类

第二步,在 stdafx.h 文件中添加头文件 afxdb.h 和其他辅助头文件(图 10-12)。

```
#include <afxcontrolbars.h>

#include <afxdb.h>
#include <tuple>
#include <map>
#include <list>
```

图 10-12 在 stdafx.h 文件中添加头文件

第三步,在应用程序类中添加 CDatabase 成员变量 m_Database 和相应函数声明(图 10-13)。

```
class CUniAppOracleODBCApp : public CWinApp
{
protected:
    CDatabase m_Database;
public :
    inline CDatabase* GetDatabase() {
        return &m_Database;
    }
    BOOL OpenDatabae();//打开数据库连接
    //从数据库中读取记录,返回记录个数
    int ReadRecords(std::list< std::pair<CString, CString> > & rs);
```

图 10-13 添加 CDatabase 成员变量和函数

第四步，编码实现 OpenDatabase() 函数，主要是调用 CDatabase 连接并打开数据库。

```
BOOL CUniAppOracleODBCApp::OpenDatabae(){
    CString strDBSource=_T("UNIVERSITY_PDBORCL");
    CString strUserName=_T("UNIVERSITY");
    CString strPassword=_T("cug");
    CString strConnect;
    strConnect.Format(
        _T("DSN=%s;UID=%s;PWD=%s"),
        strDBSource,
        strUserName,
        strPassword);

    TRY {
        m_Database.OpenEx(strConnect,CDatabase::noOdbcDialog);
    }
    CATCH(CDBException,ex1){
        AfxMessageBox(ex1->m_strError);
        AfxMessageBox(ex1->m_strStateNativeOrigin);
        return FALSE;
    }
    AND_CATCH(CMemoryException,ex2){
        ex2->ReportError();
        AfxMessageBox(_T("Memory Exception"));
        return FALSE;
    }
    AND_CATCH(CException,ex3){
        TCHAR szError[256];
        ex3->GetErrorMessage(szError,256);
        AfxMessageBox(szError);
        return FALSE;
    }
    END_CATCH
    return TRUE;
}
```

第五步，编码实现 ReadRecords() 函数，从数据库中查询返回记录，结果存放在一个列表中。

```
int CUniAppOracleODBCApp::ReadRecords(
                    std::list< std::pair<CString,CString> > & rs){
```

```
CString strSQL=
    _T("SELECT c.Cname,s.Grade FROM Course c,SC s WHERE c.Cno=s.Cno");
int rc=-1;
TRY{
    CRecordset record(GetDatabase());
    record.Open(AFX_DB_USE_DEFAULT_TYPE,strSQL,0);
    rc=record.GetRecordCount();
    CODBCFieldInfo f1,f2;
    record.GetODBCFieldInfo((short)0,f1);
    record.GetODBCFieldInfo((short)1,f2);

    CDBVariant name;
    CDBVariant grade;
    while(!record.IsEOF()){
        record.MoveNext();
        record.GetFieldValue((short)0,name);
        record.GetFieldValue((short)1,grade);
        rs.push_back(std::pair<CString,CString>(
            *(name.m_pstringA),*(grade.m_pstringW)));
    }
}
CATCH(CException,e){
    TCHAR szError[100];
    e->GetErrorMessage(szError,100);
    AfxMessageBox(szError);
}
END_CATCH
return rc;
}
```

第六步,在 BOOL CUniAppOracleODBCApp::InitInstance()函数的最后返回之前调用 OpenDatabase()函数连接数据库和 ReadRecordset()函数获取查询结果。

```
BOOL CUniAppOracleODBCApp::InitInstance(){
    .........
    //数据库连接
    if(OpenDatabae()){
        std::list< std::pair<CString,CString> > rs;
        ReadRecordset(rs);
        //TODO:输出列表中的查询结果
```

```
        return TRUE;
    }
    else {
        return FALSE;
    }
}
```

第七步,在 int CUniAppOracleODBCApp::ExitInstance()开始位置,关闭数据库连接(图 10-14)。

```
int CUniAppOracleODBCApp::ExitInstance()
{
    m_Database.Close();
    AfxOleTerm(FALSE);

    return CWinApp::ExitInstance();
}
```

图 10-14 关闭数据库

(2)采用 C++开发一个基于 OCI 的数据库应用程序,实现以下功能:输入一个系号,输出该系的系名(要求采用带参数的 SQL 语句)。

(a)按照图 10-6~图 10-11 所示,生成 MFC 文档工程 UniAppOracleOCI。

(b)配置 UniAppOracleOCI,加入 OCI 相关库。主要分为以下三步。

第一步,添加附加的头文件目录,如图 10-15 所示。

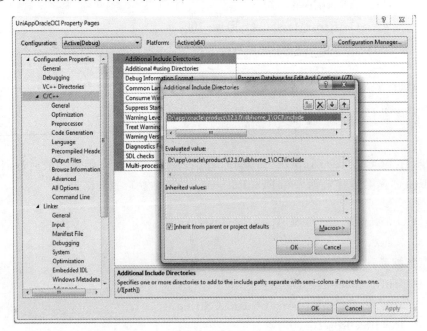

图 10-15 添加附加的头文件目录

第二步,添加附加的库目录,如图 10-16 所示。

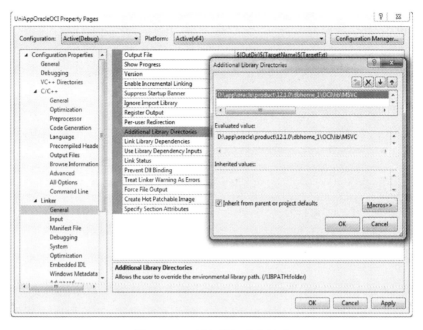

图 10-16 添加附加的库目录

第三步,添加 oci.lib 和 ociw32.lib 库文件,如图 10-17 所示。
第四步,在 stdafx.h 文件中添加♯include<oci.h>。

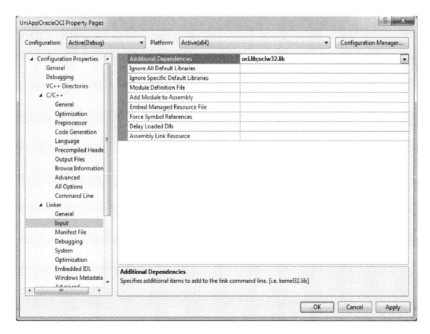

图 10-17 添加 OCI 库文件

(c)进行 OCI 编程,主要包括 OCI 环境初始化、连接数据库、读取数据进行处理、关闭数据库连接、释放 OCI 资源。一般步骤如下。

第一步,声明、创建环境句柄。

```
OCIEnv          * envhp;
OCIError        * errhp;
OCISvcCtx       * svchp;
OCIServer       * srvhp;
OCISession      * usrhp;
OCIStmt         * stmthp;
OCIInitialize(
    (ub4)(OCI_DEFAULT),
    NULL,NULL,NULL,NULL);
OCIEnvInit(&envhp,(ub4)OCI_DEFAULT,(size_t)0,NULL);
```

对于函数 OCIInitialize,第一个参数是初始化的方式,有效的初始化方式包括:OCI_DEFAULT——默认方式和 OCI_THREADED——线程环境方式,在这种方式下面内部数据结构将被保护,不会让其他线程访问。OCI_OBJECT 表示使用对象方式,OCI_EVENTS 表示使用公共订阅通知方式。

第二步,声明、创建其他句柄(错误报告句柄、服务器内容句柄、服务内容句柄)。

```
OCIHandleAlloc((dvoid *)envhp,(dvoid * *)&errhp,(ub4)OCI_HTYPE_ERROR,
    (size_t)0,(dvoid * *)0);
OCIHandleAlloc((dvoid *)envhp,(dvoid * *)&srvhp,(ub4)OCI_HTYPE_SERVER,
    (size_t)0,(dvoid * *)0);
OCIServerAttach(srvhp,errhp,(text *)dbname,(sb4)strlen(dbname),
    (ub4)OCI_DEFAULT);
OCIHandleAlloc((dvoid *)envhp,(dvoid * *)&svchp,(ub4)OCI_HTYPE_SVCCTX,
    (size_t)0,(dvoid * *)0);
OCIAttrSet((dvoid *)svchp,(ub4)OCI_HTYPE_SVCCTX,(dvoid *)srvhp,
    (ub4)0,(ub4)OCI_ATTR_SERVER,errhp);
```

第三步,创建用户句柄,设置用户名称和密码。

```
OCIHandleAlloc((dvoid *)envhp,(dvoid * *)&usrhp,
    (ub4)OCI_HTYPE_SESSION,
    (size_t)0,(dvoid * *)0);
OCIAttrSet((dvoid *)usrhp,(ub4)OCI_HTYPE_SESSION,
    (dvoid *)username,(ub4)strlen(username),
    (ub4)OCI_ATTR_USERNAME,errhp);
```

```
OCIAttrSet((dvoid *)usrhp,(ub4)OCI_HTYPE_SESSION,
    (dvoid *)password,(ub4)strlen(password),
    (ub4)OCI_ATTR_PASSWORD,errhp);
```

第四步,开始会话,构建语句句柄。

```
/* session begins */
OCISessionBegin(svchp,errhp,usrhp,
    OCI_CRED_RDBMS,OCI_DEFAULT);
OCIAttrSet((dvoid *)svchp,(ub4)OCI_HTYPE_SVCCTX,
    (dvoid *)usrhp,(ub4)0,
    (ub4)OCI_ATTR_SESSION,errhp);
/* initialize stmthp */
OCIHandleAlloc((dvoid *)envhp,(dvoid **)&stmthp,
    (ub4)OCI_HTYPE_STMT,(size_t)0,(dvoid **)0);
```

第五步,准备 SQL 语句。

```
sword status;
OCIStmtPrepare(
    stmthp,errhp,(text *)sql,
    (ub4)strlen(sql),
    (ub4)OCI_NTV_SYNTAX,(ub4)OCI_DEFAULT);
```

第六步,占位符绑定参数。

```
OCIBind *bndp=(OCIBind *)0;
int dno=14;
/* Bind the placeholder in the "Dno" statement. */
OCIBindByPos(stmthp,&bndp,errhp,1,
    (dvoid *)&dno,(sword)sizeof(dno),,SQLT_INT,
    (dvoid *)0,(ub2 *)0,(ub2 *)0,(ub4)0,(ub4 *)0,OCI_DEFAULT);
```

第七步,定义 Select 语句中的输出变量。

```
int nDname=50;
text *Dname=(text *)malloc((size_t)nDname + 1);
OCIDefine *defnp=(OCIDefine *)0;
/* Define the output variable for the select-list. */
OCIDefineByPos(stmthp,&defnp,errhp,1,
```

(dvoid *)Dname,nDname + 1,SQLT_STR,
(dvoid *)0,(ub2 *)0,(ub2 *)0,OCI_DEFAULT);

第八步,输出或处理查询返回结果。
OCIStmtExecute(svchp,stmthp,errhp,(ub4)1,(ub4)0,
　　(CONST OCISnapshot *)NULL,(OCISnapshot *)NULL,OCI_DEFAULT);
printf((const char *)Dname);

第九步,释放 OCI 资源。

/* finalize stmthp */
OCIHandleFree((dvoid *)stmthp,(ub4)OCI_HTYPE_STMT);
/* session ends */
OCISessionEnd(svchp,errhp,usrhp,(ub4)OCI_DEFAULT);
OCIServerDetach(srvhp,errhp,(ub4)OCI_DEFAULT);
/* finalize svchp,srvhp,and errhp */
OCIHandleFree((dvoid *)svchp,(ub4)OCI_HTYPE_SVCCTX);
OCIHandleFree((dvoid *)srvhp,(ub4)OCI_HTYPE_SERVER);
OCIHandleFree((dvoid *)errhp,(ub4)OCI_HTYPE_ERROR);

上述代码都在 UniAppOracleOCIApp 的头文件和实现文件中。在 CUniAppOracleO-CIApp 的声明部分定义了相关变量和函数。

```
class CUniAppOracleOCIApp:public CWinApp{
    OCIEnv      * envhp;
    OCIError    * errhp;
    OCISvcCtx   * svchp;
    OCIServer   * srvhp;
    OCISession  * usrhp;
    OCIStmt     * stmthp;
    //初始化 Oracle OCI 环境,并连接数据库服务器
    void InitialOracleOCI(const char * dbname,
        const char * username,const char * password);
    //构建语句,执行查询,返回结果并处理
    void ExecuteStatement(
        const char * sql=
        "SELECT Dname FROM Department WHERE Dno=:1");
    //释放 OCI 资源
    void UninitialOracleOCI();
    //检查错误与状态
```

```
    void CheckError(OCIError * errhp,sword status);
public：
    .........
}
```

函数 InitialOracleOCI() 的主要功能是初始化 Oracle OCI 环境,并连接数据库服务器,具体实现如下：

```
void CUniAppOracleOCIApp::InitialOracleOCI(
    const char * dbname,const char * username,const char * password){
    //1)声明、创建环境句柄
    OCIInitialize(
        (ub4)(OCI_DEFAULT),
        NULL,NULL,NULL,NULL);
    OCIEnvInit(&envhp,(ub4)OCI_DEFAULT,(size_t)0,NULL);
    //2)声明、创建其他句柄(错误报告句柄、服务器内容句柄、服务内容句柄)
    OCIHandleAlloc((dvoid * )envhp,(dvoid * * )&errhp,(ub4)OCI_HTYPE_ERROR,
        (size_t)0,(dvoid * * )0);

    OCIHandleAlloc((dvoid * )envhp,(dvoid * * )&srvhp,(ub4)OCI_HTYPE_SERVER,
        (size_t)0,(dvoid * * )0);
    OCIServerAttach(srvhp,errhp,(text * )dbname,(sb4)strlen(dbname),
        (ub4)OCI_DEFAULT);

    OCIHandleAlloc((dvoid * )envhp,(dvoid * * )&svchp,(ub4)OCI_HTYPE_SVCCTX,
        (size_t)0,(dvoid * * )0);
    OCIAttrSet((dvoid * )svchp,(ub4)OCI_HTYPE_SVCCTX,(dvoid * )srvhp,(ub4)0,
        (ub4)OCI_ATTR_SERVER,errhp);
    //3)创建用户句柄,设置用户名称和密码
    OCIHandleAlloc((dvoid * )envhp,(dvoid * * )&usrhp,(ub4)OCI_HTYPE_SESSION,
        (size_t)0,(dvoid * * )0);
    OCIAttrSet((dvoid * )usrhp,(ub4)OCI_HTYPE_SESSION,
        (dvoid * )username,(ub4)strlen(username),
        (ub4)OCI_ATTR_USERNAME,errhp);
    OCIAttrSet((dvoid * )usrhp,(ub4)OCI_HTYPE_SESSION,
        (dvoid * )password,(ub4)strlen(password),
        (ub4)OCI_ATTR_PASSWORD,errhp);

    //4)开始会话,构建语句句柄
    / * session begins * /
```

```
    OCISessionBegin(svchp,errhp,usrhp,
        OCI_CRED_RDBMS,OCI_DEFAULT);
    OCIAttrSet((dvoid *)svchp,(ub4)OCI_HTYPE_SVCCTX,
        (dvoid *)usrhp,(ub4)0,
        (ub4)OCI_ATTR_SESSION,errhp);

    /* initialize stmthp */
    OCIHandleAlloc((dvoid *)envhp,(dvoid **)&stmthp,
        (ub4)OCI_HTYPE_STMT,(size_t)0,(dvoid **)0);

}
```

函数 UninitialOracleOCI() 的主要功能是关闭数据库服务器连接,反初始化 Oracle OCI 环境,释放之前申请的 OCI 资源,具体实现如下:

```
void CUniAppOracleOCIApp::UninitialOracleOCI(){
    /* finalize stmthp */
    OCIHandleFree((dvoid *)stmthp,(ub4)OCI_HTYPE_STMT);

    /* session ends */
    OCISessionEnd(svchp,errhp,usrhp,(ub4)OCI_DEFAULT);
    OCIServerDetach(srvhp,errhp,(ub4)OCI_DEFAULT);

    /* finalize svchp,srvhp,and errhp */
    OCIHandleFree((dvoid *)svchp,(ub4)OCI_HTYPE_SVCCTX);
    OCIHandleFree((dvoid *)srvhp,(ub4)OCI_HTYPE_SERVER);
    OCIHandleFree((dvoid *)errhp,(ub4)OCI_HTYPE_ERROR);
}
```

函数 CheckError() 的主要功能是检查 OCI 执行状态是否出错,具体实现如下:

```
void CUniAppOracleOCIApp::CheckError(OCIError *errhp,sword status)
{
    text errbuf[512];
    sb4 errcode=0;

    switch(status)
    {
    case OCI_SUCCESS:
        break;
    case OCI_SUCCESS_WITH_INFO:
```

```
            fprintf(stderr,"OCI_SUCCESS_WITH_INFO\n");
            break;
        case OCI_ERROR:
            OCIErrorGet((dvoid *)errhp,(ub4)1,(text *)NULL,&errcode,
                errbuf,(ub4)sizeof(errbuf),OCI_HTYPE_ERROR);
            fprintf(stderr,"%.*s\n",512,errbuf);
            break;
        case OCI_NEED_DATA:
            fprintf(stderr,"OCI_NEED_DATA\n");
            break;
        case OCI_NO_DATA:
            fprintf(stderr,"OCI_NO_DATA\n");
            break;
        case OCI_INVALID_HANDLE:
            fprintf(stderr,"OCI_INVALID_HANDLE\n");
            break;
        case OCI_STILL_EXECUTING:
            fprintf(stderr,"OCI_STILL_EXECUTING\n");
            break;
        case OCI_CONTINUE:
            fprintf(stderr,"OCI_CONTINUE\n");
            break;
        default:
            break;
    }

    if(status!=OCI_SUCCESS && status!=OCI_SUCCESS_WITH_INFO)
            UninitialOracleOCI();
}
```

函数 ExecuteStatement() 的主要功能执行 SQL 语句，并获取返回值，具体实现见 CUniAppOracleOCIApp 中的对应函数。

```
    void CUniAppOracleOCIApp::ExecuteStatement(const char * sql){
        /*
        create table Department(
            Dno number(10),
            Dname varchar2(50),
            Daddress varchar2(50),
            primary key(Dno)
```

);
sql="SELECT Dname FROM Department WHERE Dno=:1"
*/
sword status;
OCIStmtPrepare(
    stmthp,errhp,(text *)sql,
    (ub4)strlen(sql),
    (ub4)OCI_NTV_SYNTAX,(ub4)OCI_DEFAULT);

OCIBind * bndp=(OCIBind *)0;
int dno=14;
/* Bind the placeholder in the "Dno" statement. */
if(status=OCIBindByPos(stmthp,&bndp,errhp,1,
    (dvoid *)&dno,(sword)sizeof(dno),SQLT_INT,
    (dvoid *)0,(ub2 *)0,(ub2 *)0,(ub4)0,(ub4 *)0,OCI_DEFAULT))
{
    CheckError(errhp,status);
    return;
}

/* Allocate the dept buffer now that you have length. */
/* the deptlen should eventually get from dschndl3.   */
int nDname=50;
text * Dname=(text *)malloc((size_t)nDname + 1);
OCIDefine * defnp=(OCIDefine *)0;
/* Define the output variable for the select-list. */
if(status=OCIDefineByPos(stmthp,&defnp,errhp,1,
    (dvoid *)Dname,nDname + 1,SQLT_STR,
    (dvoid *)0,(ub2 *)0,(ub2 *)0,OCI_DEFAULT))
{
    CheckError(errhp,status);
    return;
}

/*
 * Prompt for the employee's department number,and verify
 * that the entered department number is valid
 * by executing and fetching.
 */

```
        if((status=OCIStmtExecute(svchp,stmthp,errhp,(ub4)1,(ub4)0,
          (CONST OCISnapshot *)NULL,(OCISnapshot *)NULL,OCI_DEFAULT))
           &&(status！=OCI_NO_DATA))
    {
            CheckError(errhp,status);
            return;
    }
    if(status==OCI_NO_DATA)
            printf("The dept you entered doesn't exist. \n");
    else
            printf((const char *)Dname);
}
```

在InitInstance()函数的结尾处调用OCI相关函数,连接数据库,获取数据并处理,具体实现如下:

```
// CUniAppOracleOCIApp initialization
BOOL CUniAppOracleOCIApp::InitInstance(){
    ………
    m_pMainWnd->DragAcceptFiles();
    const char * dbname="pdborcl";
    const char * usr="university";
    const char * psw="cug";
    InitialOracleOCI(dbname,usr,psw);
    ExecuteStatement();
    return TRUE;
}
```

在ExitInstance()函数的开始处调用OCI相关函数,关闭连接数据库,释放OCI中申请的一些句柄资源,具体实现如下:

```
int CUniAppOracleOCIApp::ExitInstance(){
    UninitialOracleOCI();
    AfxOleTerm(FALSE);
    return CWinApp::ExitInstance();
}
```

(3)配置Oracle数据库的ODBC连接,采用C++开发一个基于ODBC的数据库应用程序(控制台应用程序),输出所有学生姓名,并插入一条有效的学生选课记录(实验环境为:

WinXP 32Bits + VC6.0+ Oracle 11g Express)。

(a)配置 UNIVERSITY 数据库的 ODBC 数据源。

第一步,在 Windows 中找到数据源(ODBC)菜单,如图 10-18 所示。弹出对话框,选择"驱动程序"选项卡,查看是否有"Oracle in XE",如图 10-19 所示。

图 10-18 Windows 中的 ODBC 数据源菜单

图 10-19 Oracle in XE 驱动程序

第二步,如图 10-20 所示,选择"用户 DSN",点击"添加(D)..."按钮,弹出如图 10-21 所示界面。选择"Oracle in XE"驱动程序,点击"完成",弹出如图 10-22 所示界面。在 Data Source Name、TNS Service Name、User ID 等中输入信息,点击"Test Connection"按钮,输入用户密码(图 10-23),完成连接测试。最后,完成数据源 UNIVERSITY 配置,如图 10-24 所示。

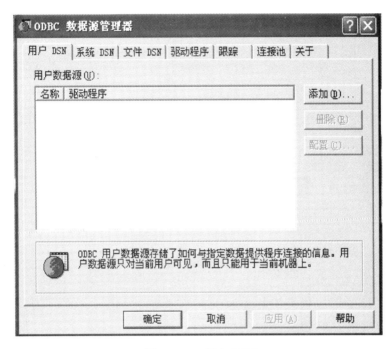

图 10-20 用户 DSN

图 10-21 选择数据源驱动程序

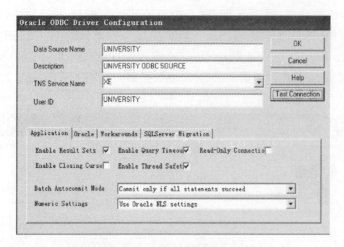

图 10-22 输入数据源配置信息

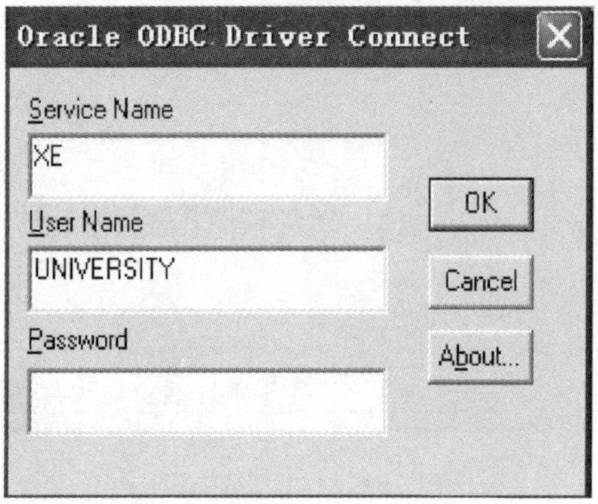

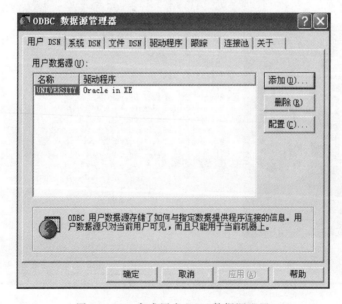

图 10-23 输入数据源的用户密码

图 10-24 完成用户 DSN 数据源配置

(b) 基于 ODBC 开发数据库应用程序(控制台应用程序)。

第一步,建立 Visual Studio 6.0 的 VC 控制台应用程序,如图 10-25、图 10-26 所示。

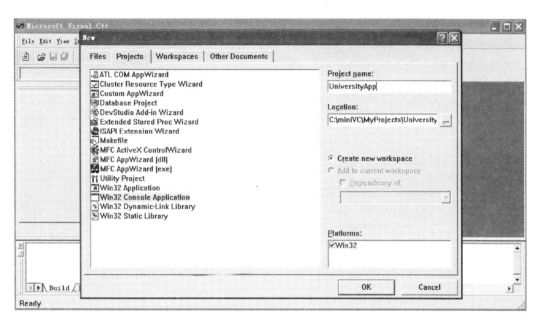

图 10-25 选择 Win32 Console Application 工程类型

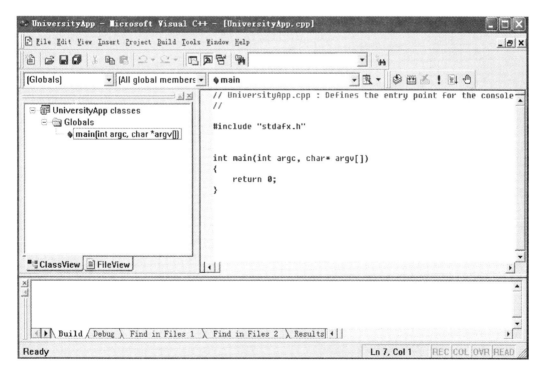

图 10-26 生成 Win32 Console Application 简单工程

第二步,在 UniversityApp.cpp 中添加相关头文件和全局变量。

```
#include "stdio.h"
#include "stdlib.h"
#include <Windows.h>
#include "sql.h"
#include "sqlext.h"
#include "sqltypes.h"
#define   STR_LEN 50

HENV          henv;         /* environment handle */
HDBC          hdbc;         /* connection handle  */
SQLHSTMT      hstmt;        /* statement handle   */
SDWORD        retcode;      /* return code        */
```

第三步,在 UniversityApp.cpp 中添加 ODBC 环境初始化函数。

```
/**
   initial ODBC,and connect to the data source
*/
bool initial(SQLCHAR * odbc,SQLCHAR * userid,SQLCHAR * password){
    SQLCHAR      SqlState[6],Msg[SQL_MAX_MESSAGE_LENGTH];
    SQLINTEGER   NativeError;
    SQLSMALLINT  MsgLen;

    retcode=SQLAllocEnv(&henv);
    retcode=SQLSetEnvAttr(henv,
        SQL_ATTR_ODBC_VERSION,(void *)SQL_OV_ODBC3,
        SQL_IS_INTEGER);
    retcode=SQLAllocConnect(henv,&hdbc);
    retcode=SQLConnect(hdbc,odbc,SQL_NTS,userid,
        SQL_NTS,password,SQL_NTS);
      if(retcode!=SQL_SUCCESS){
        if((retcode=SQLGetDiagRec(SQL_HANDLE_DBC,
            hdbc,1,SqlState,
            &NativeError,Msg,sizeof(Msg),
            &MsgLen))!=SQL_NO_DATA)
        printf("SqlState=%s\n Message=%s\n",SqlState,Msg);
        return false;
    }
```

```
        else
            return true;
}
```

第四步,在 UniversityApp.cpp 中添加 ODBC 环境清理函数。

```
/**
   release all the handes created by the initial function
*/
void uninitial(){
    SQLDisconnect(hdbc);
    SQLFreeConnect(hdbc);
    SQLFreeEnv(henv);
}
```

第五步,在 UniversityApp.cpp 中添加学生姓名查询函数。

```
/**
query all the students' names,and output them
*/
bool executeQuery(){
    SQLCHAR        sname[STR_LEN]="";
      #ifdef _WIN64
            SQLLEN snind;
      #elif BUILD_REAL_64_BIT_MODE
            SQLLEN snind;
      #else
            SQLINTEGER snind;
      #endif

        retcode=SQLAllocHandle(SQL_HANDLE_STMT,hdbc,&hstmt);
        if(retcode! =SQL_SUCCESS)return false;

        SQLBindCol(hstmt,1,SQL_C_CHAR,sname,sizeof(sname),&snind);

        SQLExecDirect(hstmt,
            (SQLCHAR * )"select sname from student",SQL_NTS);

        while((retcode=SQLFetch(hstmt))! =SQL_NO_DATA){
            if(snind==SQL_NULL_DATA)
```

```
                printf("NULL  \n");
            else
                printf("SNAME=%s \n",sname);
        }
        SQLFreeStmt(hstmt,SQL_CLOSE);
        return true;
}
```

第六步,在 UniversityApp.cpp 中添加学生记录添加函数。

```
    / * *
    insert a sc record
*/
bool executeInsertion(){

    SQLINTEGER tno,cno;
    SQLCHAR site[STR_LEN];

    #ifdef _WIN64
        SQLLEN tnoInd=0,cnoInd=0,siteInd=0;
    #elif   BUILD_REAL_64_BIT_MODE
        SQLLEN tnoInd=0,cnoInd=0,siteInd=0;
    #else
        SQLINTEGER tnoInd=0,cnoInd=0,siteInd=0;
    #endif

    SQLCHAR          SqlState[6],Msg[SQL_MAX_MESSAGE_LENGTH];
    SQLINTEGER       NativeError;
    SQLSMALLINT      MsgLen;

    retcode=SQLAllocHandle(SQL_HANDLE_STMT,hdbc,&hstmt);
    if(retcode!=SQL_SUCCESS)
        return false;

    retcode=SQLPrepare(hstmt,(SQLCHAR*)"insert into tc(tno,cno,site) values(?,?,?)",SQL_NTS);
        if(retcode!=SQL_SUCCESS){
            retcode=SQLGetDiagRec(SQL_HANDLE_STMT,hstmt,1,
                SqlState,&NativeError,Msg,sizeof(Msg),&MsgLen);
            printf("SqlState=%s\n Message=%s\n",SqlState,Msg);
```

```
        return false;
    }

    if((retcode=SQLBindParameter(hstmt,1,SQL_PARAM_INPUT,SQL_C_ULONG,
        SQL_INTEGER,4,0,&tno,sizeof(tno),
        &tnoInd))!=SQL_SUCCESS)return false;
    if((retcode=SQLBindParameter(hstmt,2,SQL_PARAM_INPUT,SQL_C_ULONG,
        SQL_INTEGER,4,0,&cno,sizeof(cno),
        &cnoInd))!=SQL_SUCCESS)return false;
    if((retcode=SQLBindParameter(hstmt,3,SQL_PARAM_INPUT,SQL_C_CHAR,
        SQL_INTEGER,0,0,&site[0],50,
        &siteInd))!=SQL_SUCCESS)return false;

    tno=1;
    cno=10;
    site[0]='J';
    site[1]='2';
    site[3]='0';
    retcode=SQLExecute(hstmt);
    if(retcode!=SQL_SUCCESS){
        retcode=SQLGetDiagRec(SQL_HANDLE_STMT,hstmt,1,
            SqlState,&NativeError,Msg,sizeof(Msg),&MsgLen);
        printf("SqlState=%s\n Message=%s\n",SqlState,Msg);
        return false;
    }

    SQLEndTran(SQL_HANDLE_ENV,henv,SQL_COMMIT);
    printf("Insertion is successful\n");

    SQLFreeStmt(hstmt,SQL_CLOSE);

    return true;
}
```

10.4 实验报告

实验报告按照附录1的格式进行编写(斜体字表示学生实验报告中要编写或要填写的内容)。

11 数据库应用开发(Java)

采用高级语言 Java 进行数据库应用程序开发,一般都会采用 JDBC(Java DataBase Connectivity)进行数据库链接。JDBC 是 Oracle 公司给出的一个 Java 语言访问关系数据库的 Java 接口标准,这个标准基于 X/Open SQL Call Level Interface ,并与 SQL 标准兼容。其体系结构如图 11-1 所示,主要包括 JDBC API、JDBC Driver Manager 和 JDBC Driver 三个部分。各数据库供应商可以在他们的驱动程序中实现并扩展这个接口。相应的驱动程序称为 JDBC 驱动程序。一般厂家在实现 JDBC 接口,设计自己的 JDBC 驱动程序时,都有个性化的处理。如 Oracle 的 JDBC 驱动程序除了支持标准的 JDBC API,还支持 Oracle 特定的数据类型以提高性能。JDBC API 是 Java 编程语言和各种数据库之间独立于数据库的连接的行业标准,它为基于 SQL 的数据库访问提供了一个调用级 API。JDBC API 可以做三件事:①建立与数据库的连接或访问任何表格数据源;②发送 SQL 语句;③处理结果。

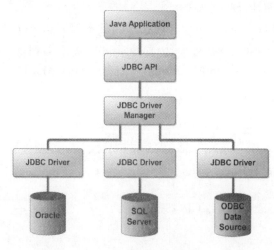

图 11-1 JDBC 体系结构

11.1 实验目的

(1)掌握基于 JDBC 的编程原理与基本步骤。
(2)掌握基于 Oracle JDBC 的三种常见数据库连接方式,并能进行简单数据库应用程序开发。

11.2 实验平台

(1)操作系统:Windows XP、Windows Server 2003 及后续版本、Windows 7 及后续版本。
(2)数据库管理系统:Oracle 11g 及后续版本。
(3)Java 开发环境:Java SDK 7 及以上版本、JetBrains IDEA 或 Eclipse。
(4)本实验以第 2、第 3 章构建的 UNIVERSITY 数据库为例进行实验。

11.3 实验内容

(1)采用 Java 开发一个基于 JDBC 的数据库应用程序,实现以下功能:输出每个学生选修课程总成绩。

(a)基于 IDEA 构建 Java 工程。在 IDEA 中新建工程,选择"Java"类型,设置 Project SDK (图 11-2);然后选择工程模版,如图 11-3 所示;接着,输入工程名称和存放位置,如图 11-4 所示;最后生成如图 11-5 所示的 Hello World 工程。

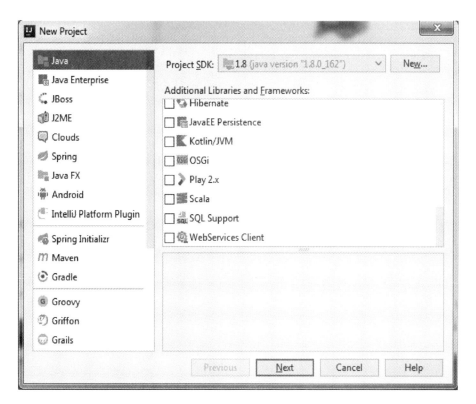

图 11-2 IDEA 构建 Java 工程

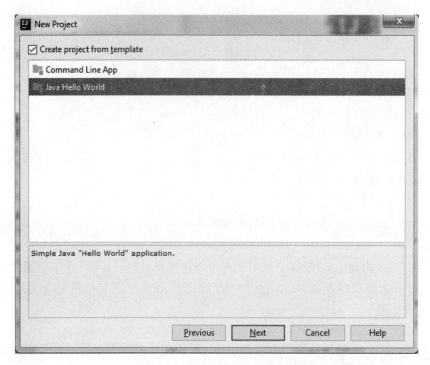

图 11-3　IDEA Java 工程模板

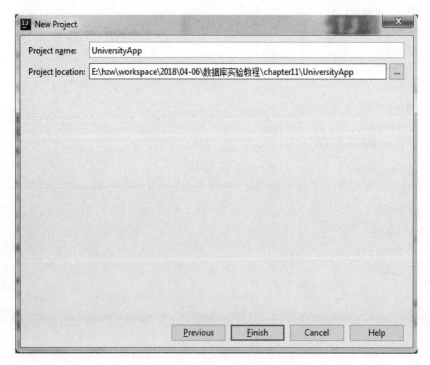

图 11-4　IDEA Java 工程基本信息

11 数据库应用开发(Java)

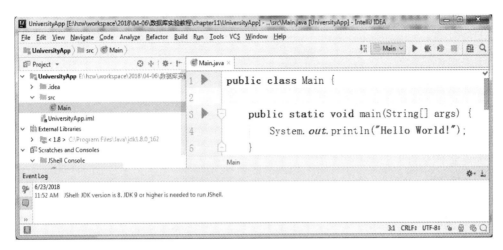

图 11-5 IDEA Java HelloWorld

(b)添加依赖的 Jar 包、ojdbc8.lib 等。点击"File ->Project Structure ->Libraries",再点击加号,添加 ojdbc8.jar 和其他相关的依赖 Jar 包(图 11-6~图 11-8)。

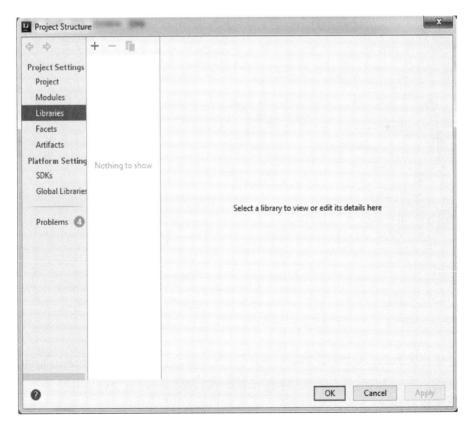

图 11-6 IDEA Java 工程添加依赖库

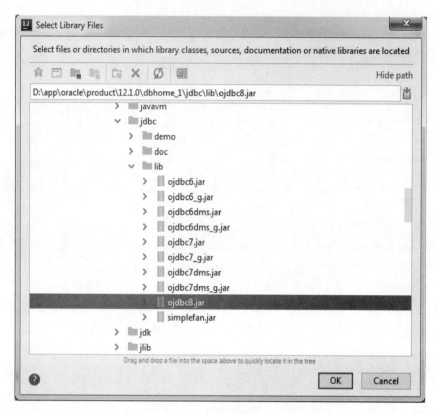

图 11-7　IDEA Java 工程添加 JDBC 下的依赖库

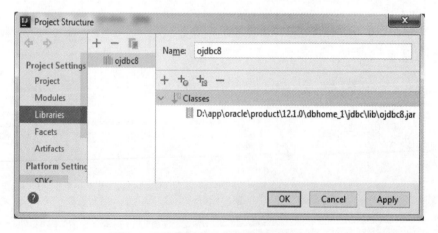

图 11-8　IDEA Java 工程添加 ojdbc8.jar

　　(c)编写 Java 程序。采用 Java 语言,基于 JDBC 编写数据库应用程序,与其他程序的主要区别在于增加了连接数据库的内容。对于 Oracle 12c,我们一般连接的是其中的一个 PLUG-GABLE 数据库,例如默认的 PDBORCL。如果是 Oracle 12c 之前的版本,则不存在可插拔数据库的问题。但不管是否是可插拔数据库,一般而言,通过 JDBC 连接 Oracle 数据库基本程序都是一样的。不同的是,URL 有以下三种格式。

格式一：Oracle JDBC Thin using an SID：
jdbc:oracle:thin:@host:port:SID
例如：jdbc:oracle:thin:@localhost:1521:orcl
格式二：Oracle JDBC Thin using a ServiceName：
jdbc:oracle:thin:@//host:port/service_name
例如：jdbc:oracle:thin:@//localhost:1521/pdborcl
格式三：Oracle JDBC Thin using a TNSName：
jdbc:oracle:thin:@TNSName
例如：jdbc:oracle:thin:@TNS_ALIAS_NAME

在上面三种格式中，格式二是比较常用的。需要注意的是，这里的格式"@"后面有"//"，"port"后面的":"换成了"/"。这种格式是 Oracle 推荐的格式，因为对于集群来说，每个节点的 SID 可能是不一样的，但是 SERVICE_NAME 可以包含所有节点。连接本书示例数据库 UNIVERSITY 的代码如下：

```java
public static Connection getConnection(){
    Connection conn=null;
    try {
        Class.forName("oracle.jdbc.driver.OracleDriver");//找到 oracle 驱动器所在的类
        String url="jdbc:oracle:thin:@//localhost:1521/pdborcl"; //URL 地址
        String username="university";
        String password="cug";
        conn=DriverManager.getConnection(url,username,password);
    } catch(ClassNotFoundException e){
        e.printStackTrace();
    } catch(SQLException e){
        e.printStackTrace();
    }
    return conn;
}
```

获取连接之后，就可以构建语句，执行并获取查询结果。

```java
public static void main(String[] args){
    String sql="select sno,sum(grade)sg from sc group by sno";
    Connection connection=getConnection();
    try {
        Statement statement=connection.createStatement();
        ResultSet resultSet=   statement.executeQuery(sql);
```

```
            while(resultSet.next()){
                String sno=resultSet.getString("sno");
                double sg=resultSet.getDouble("sg");
                System.out.println(sno);
                System.out.println(sg);
            }
        }
        catch(SQLException e){
            e.printStackTrace();
        }
    }
```

(2)采用 Java 开发一个基于 JDBC 的数据库应用程序,实现以下功能:输出每个学生的姓名及其选修课程平均成绩。

解决本题的思路和方法与前文内容是一样的,唯一不同的地方是更换了 SQL 语句。将上一题的 SQL 语句替换成如下语句:

select student.sname,sa.ag from student,(select sno,avg(grade)from sc group by sno)as sa(sno,ag)where sa.sno=student.sno;

上面的 SQL 语句采用了基于派生表的查询,即先计算每个学号的学生的平均成绩,然后与 Student 表进行连接,得到所要的查询结果。

11.4 实验报告

实验报告按照附录 1 的格式进行编写(斜体字表示学生实验报告中要编写或要填写的内容)。

12 数据库应用开发(C♯)

本书采用高级语言 C♯ 进行数据库应用程序开发,ADO.NET 一般是最常见数据库连接程序,并且大多会和 SQL Server 进行搭配组合。如果数据库采用 Oracle,其连接程序常用 Oracle Data Privider for .NET(ODP.NET)。ODP.NET 优化了 ADO.NET 在 Oracle 数据库连接中的数据处理的性能。有三种类型的 ODP.NET 驱动程序:ODP.NET 托管驱动、ODP.NET 非托管驱动和 ODP.NET 核心驱动。ODP.NET 托管驱动是 100% .NET 编码。相比于 ODP.NET 的托管驱动,ODP.NET 非托管驱动包含更多特性,因为它能访问构建于 Oracle 数据库客户端的功能。ODP.NET 核心驱动是指多平台的.NET 核心应用程序设计,目前可以支持 Linux 和 Windows 操作系统。

ODP.NET 需要单独安装,安装程序可以到 Oracle 的官方网站下载。下载的时候需要 Oracle 账号,下载地址:http://www.oracle.com/technetwork/topics/dotnet/downloads/index.html。以 Visual Studio 2017 和 Oracle 12c 为例,安装过程如图 12-1~图 12-4 所示。此外,Oracle 也提供了关于.NET 的一些示例工程,详细参见 https://github.com/oracle/dotnet-db-samples。

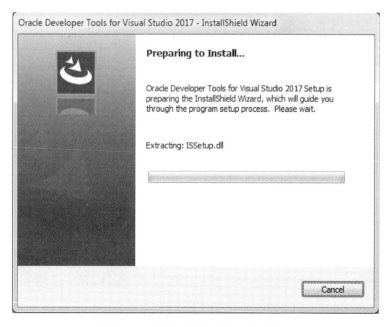

图 12-1 安装 ODTforVS2017_122011

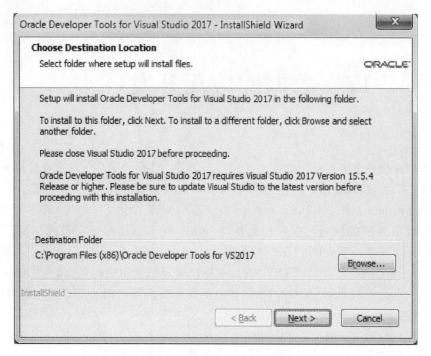

图 12-2　选择 ODTforVS 安装位置

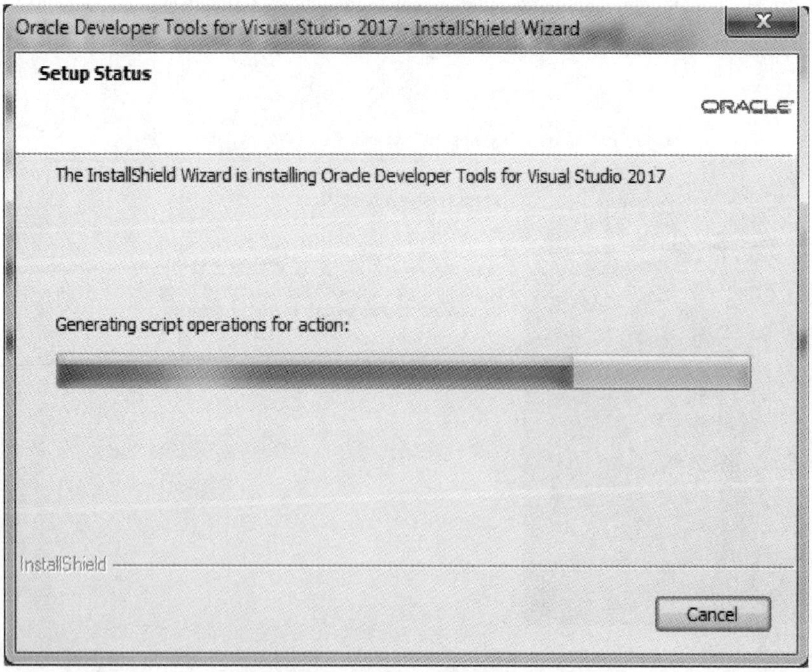

图 12-3　ODTforVS 安装过程

图 12-4 ODTforVS 安装结束

12.1 实验目的

掌握基于 C♯ 高级语言进行数据库应用程序开发的方法。

12.2 实验平台

(1)操作系统:Windows XP、Windows Server 2003 及后续版本、Windows 7 及后续版本。
(2)数据库管理系统:Oracle 11g 及后续版本。
(3).Net 开发环境:Visual Studio 2010 及其后续版本。
(4)本实验以第 2、第 3 章构建的 UNIVERSITY 数据库为例进行实验。

12.3 实验内容

(1)采用 C♯ 开发一个基于 ODP.NET 的数据库应用程序,实现以下功能:输出每个学生选修课程平均成绩。

第一步,构建 C♯ 程序,选择"Windows Form App(.NET Framework)",如图 12-5 所示。构建 UniApp 工程,然后将"Form"改为"MainForm",如图 12-6 所示。

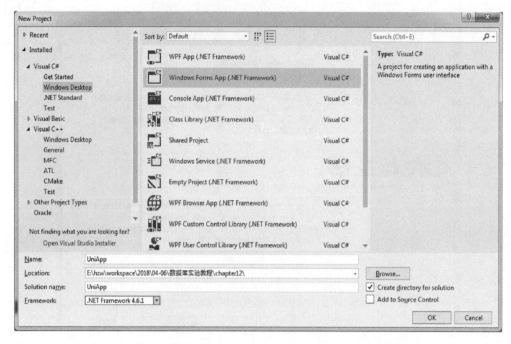

图 12-5 选择"Windows Form App(.NET Framework)"工程类型

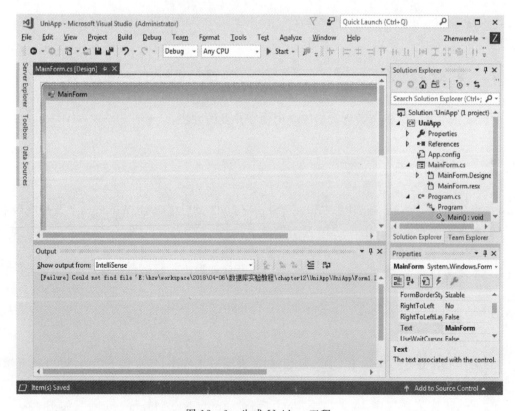

图 12-6 生成 UniApp 工程

第二步,添加引用,如图 12-7 所示,选择"Project→Add Reference…",弹出如图 12-8 所示的界面,选择"Oracle.ManagedDataAccess"。

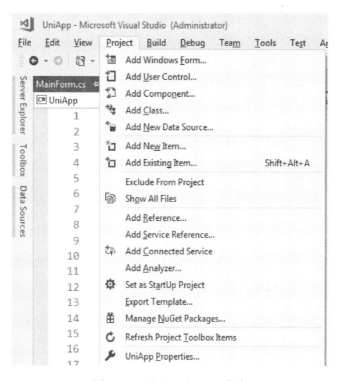

图 12-7　Add Reference 菜单

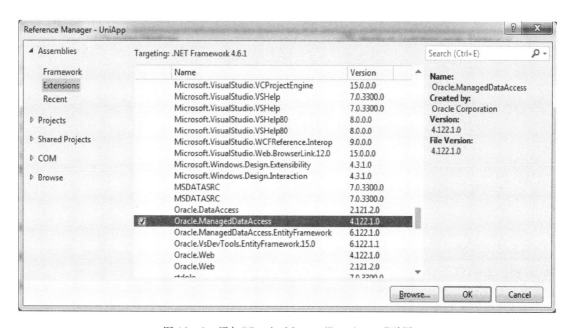

图 12-8　添加"Oracle.ManagedDataAccess"引用

第三步,在 MainFrame.cs 中添加获取连接的函数。该函数通过传入数据库服务器所在主机名或 IP 地址、监听端口、服务名称、用户名称和用户密码,构建一个 Connection 对象并打开。具体编码如下:

```
public static OracleConnection GetConnection(
        string host,string port,string serviceName,
        string user,string password)
    {
    string constr=
      "Data Source=(DESCRIPTION=(ADDRESS=(PROTOCOL=TCP)(HOST="
        +host+")(PORT="
        +port+"))(CONNECT_DATA=(SERVER=DEDICATED)(SERVICE_NAME="
        +serviceName+")));"
      +"User Id="
      +user+"; Password="
      +password+";";

    try
    {
      OracleConnection conn=new OracleConnection(constr);
      conn.ConnectionString=constr;
      conn.Open();
      return conn;
    }
    catch(Exception ex)
    {
      Console.WriteLine(ex.Message);
      Console.WriteLine(ex.Source);
    }
    return null;
  }
```

第四步,在 MainFrame.cs 中添加获取查询函数。该函数在获取数据库连接后,构建一个 Command 对象,执行获取学生平均成绩的查询语句,返回一个 DataReader 对象,最后获取返回结果。具体编码如下:

```
public static void ExecuteQuery(string sql)
  {
```

```
if(sql==null)
    sql="select sno,avg(grade)sg from sc group by sno;";

try
{
    OracleConnection conn=GetConnection("localhost",
            "1521","pdborcl","UNIVERSITY","cug");
    OracleCommand comm=new OracleCommand(sql,conn);

    OracleDataReader dr=comm.ExecuteReader();
    while(dr.Read())
    {
        long   d1=dr.GetInt64(0);
        float d2=dr.GetFloat(1);
        Console.WriteLine(d1);
        Console.WriteLine(d2);
    }
}
catch(Exception ex)
{
    Console.WriteLine(ex.Message);
    Console.WriteLine(ex.Source);
}
}
```

12.4 实验报告

实验报告按照附录1的格式进行编写(斜体字表示学生实验报告中要编写或要填写的内容)。

13 数据备份与恢复

数据库备份也称为数据库转储,方法一般分为逻辑备份和物理备份。物理备份是复制数据库的纹理文件。逻辑备份是将数据库对象的定义和数据导出到指定的文件中。按每次备份的内容是否完全可以划分为完全备份和增量备份。完全备份是备份整个数据库。增量备份则只备份上次备份以来有变化的数据。以 Oracle 数据库为例,主要有三种标准的备份方法,它们分别是导出/导入(expdp/impdp 或 exp/imp)、热备份和冷备份。导出备份是一种逻辑备份,冷备份和热备份是物理备份。与数据库备份相对应,数据库的恢复也可以分为逻辑恢复、物理恢复、增量恢复和完全恢复。

13.1 实验目的

掌握数据库逻辑备份和逻辑恢复的方法。

13.2 实验平台

(1)操作系统:Windows XP、Windows Server 2003 及后续版本、Windows 7 及后续版本。
(2)数据库管理系统:根据实际情况,自己选择 Oracle 或 SQL Server 或 MySQL 中的一种数据库管理系统软件。
(3)本实验以第 2、第 3 章构建的 UNIVERSITY 数据库为例进行实验。

13.3 实验内容

(1)逻辑备份(导出)UNIVERSITY 数据库的全部内容到指定文件中。

> expdp university/cug@localhost/pdborcl directory=DATA_FILE_DIR dumpfile=university.dmp logfile=university.log schemas=university

命令执行结果如图 13-1 所示。需要说明的是:university/cug@localhost/pdborcl 分别为用户名、用户密码、数据库服务器名称或 IP、数据库实例名称或服务名称。其中,localhost 可以省略不写。接下来的是导出目录,其中 DATA_FILE_DIR 是一个定义在 SYS 方案中的 ALL_DIRECTORIES 视图中的目录名称,如图 13-2 所示,它对应的实际目录可以人为设置。例如，本例中为 d:\app\oracle\product\12.1.0\dbhome_1\demo\schema\sales_history\。导出的 university.dmp 和 university.log 文件就存放在该目录下。schemas=university 指定了需要导出的是 UNIVERSITY 模式下的所有数据库对象，包括数据表、视图等。

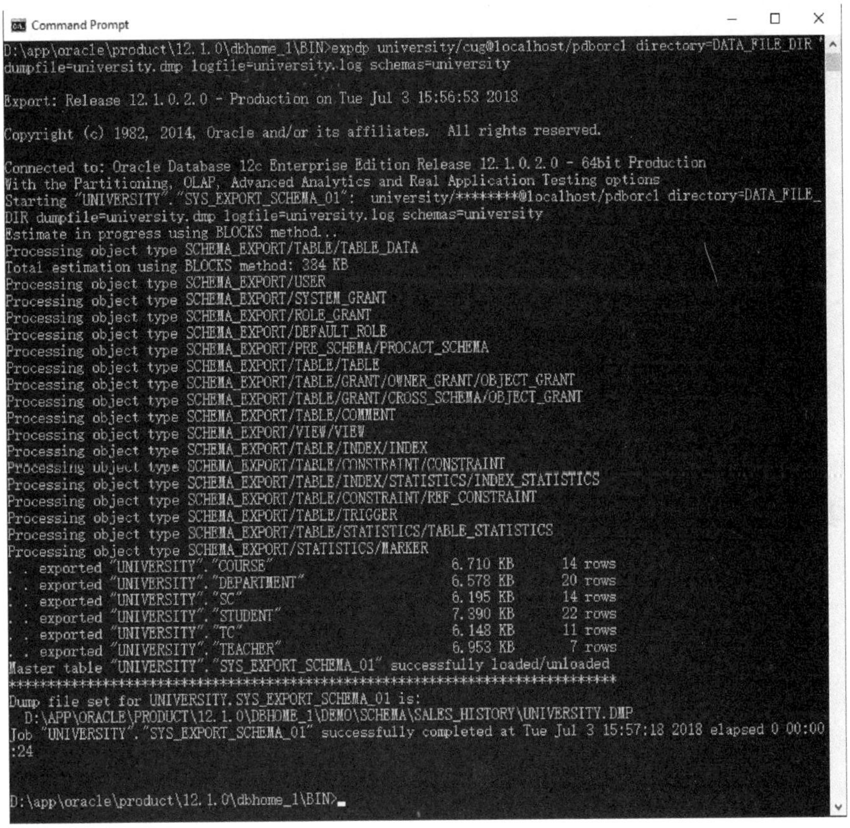

图 13-1　EXPDP 命令执行结果

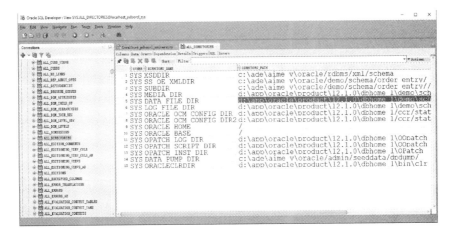

图 13-2　系统定义的目录视图

除了采用 EXPDP 命令导出外,还可以使用 EXP 命令。但 EXP 是属于低版本提供的命令,建议在 Oracle 12c 及其以后的版本中使用 EXPDP 命令。此外,除了命令行导出,SQL Developer 中也提供了数据库导出功能,如图 13-3、图 13-4 所示。在图 13-3 中,用户需要选择一个可用的数据连接,然后选择需要导出的数据库对象和内容。图 13-4 对导出的数据库对象和内容采用树形结构进行汇总展示。这样导出的数据库为一个 SQL 脚本文件。希望再次恢复数据库的时候,只需要执行该脚本文件即可。

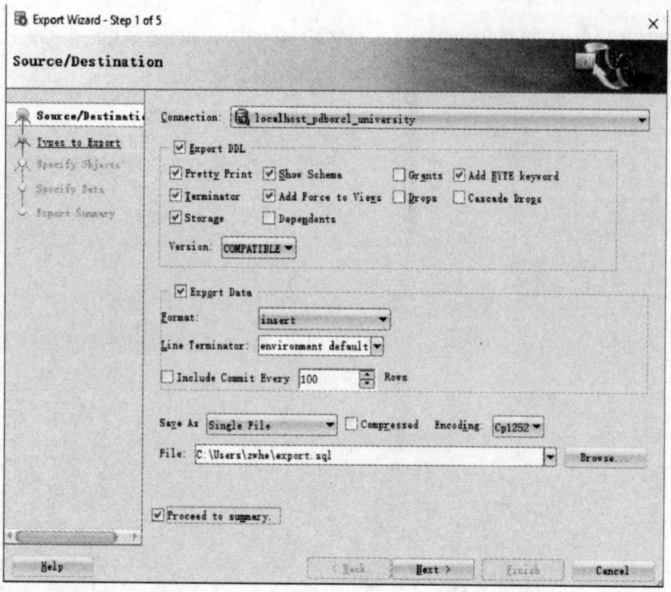

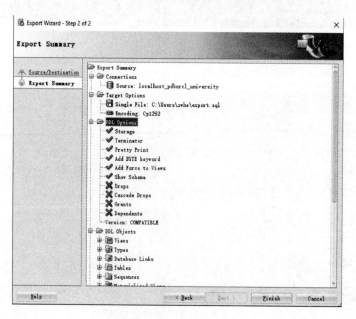

图 13-3 SQL Developer 数据库导出(选择连接和导出内容)

图 13-4 导出汇总

(2)逻辑备份(导出)UNIVERSITY 数据库中的 Student 表和 Course 表到指定文件中。

> expdp university/cug@localhost/pdborcl directory=DATA_FILE_DIR dumpfile=sc.dmp logfile=sc.log tables=(student,course)

上面的命令将指定数据表 Student 和 Course 导出到 sc.dmp 文件中，执行结果如图 13-5 所示。

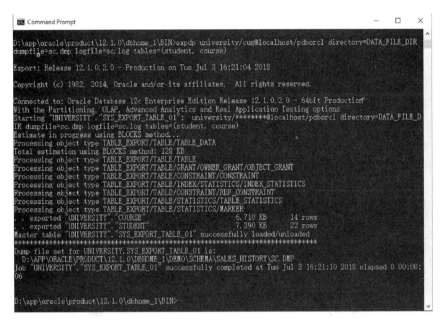

图 13-5 导出部分数据表

(3)逻辑恢复(导入)UNIVERSITY 数据库的全部内容。

> impdp university/cug@pdborcl directory=DATA_FILE_DIR dumpfile=university.dmp logfile=university.log schemas=university

说明：上面的命令可以导入整个 UNIVERSITY 数据库，前提是在数据库中 UNIVERSITY 用户下的方案为空，否则会提出数据库对象已经存在的错误。

(4)逻辑恢复(导入)UNIVERSITY 数据库中的 Student 表。

> impdp university/cug@pdborcl directory=DATA_FILE_DIR dumpfile=university.dmp logfile=university.log tables=(student,course)

13.4 实验报告

实验报告按照附录 1 的格式进行编写(斜体字表示学生实验报告中要编写或要填写的内容)。

14　综合实验(课程设计)

自选题目,设计一个数据库应用系统,按照数据库设计的步骤完成文档,包括需求分析、概念设计、逻辑设计、物理设计几个部分(可以参考第 8 章数据库设计部分内容)。然后进行功能设计,采用高级语言或脚本语言(如 Java、C/C++、C♯、Python、PHP 等)加数据库方式实现所设计的功能,形成一个数据库应用系统(可以参考第 10~12 章数据库应用开发部分内容)。要求如下。

(1)数据库结构要求:选题中实体个数必须有三个或以上,最终形成的数据表必须有五个及以上。

(2)数据库数据要求:构建的数据库中需要有足够的测试数据(可以采用随机方式生成,也可以手工输入),至少需要满足所设计功能的实现与结果展示。

(3)应用程序功能要求:选题中功能个数必须达到五个或以上,每个功能中必须采用高级语言嵌入简单 SQL 语句或调用存储过程/自定义函数这两种方式中的一种加以实现。基本的原则是让计算和数据尽可能都在服务器端,客户端负责取回计算结果。

(4)实验文档要求:形成课程设计报告,格式见附录 2。

附录1　实验报告要求

实验报告(宋体,二号,居中)

实验名称:(填写当次实验的名称)
实验环境:(填写实验所采用的软硬件配置,包括操作系统、主频、内存等)
实验者:(填写实验者姓名,采用的格式为:学号+姓名,如果有多个实验者,则每个实验者之间以分号分割,写在第一个位置的实验者为本组的组长)
实验时间:(填写实验进行的时间)

(接下来开始编写实验内容,每一个实验内容至少包括以下几个部分:内容描述、操作步骤描述和操作结果描述,以第二章的实验为例)

1. 采用 SQL 语言建立 Department 表(注意:这里采用的是 SQL Server 2008,也可以采用 MySQL 或 Oracle)

```
CREATE   TABLE Department(
Dno INT PRIMARY KEY,
Dname NVARCHAR(50),
Daddress NVARCHAR(50));
```

2. 采用 SQL Server Management Studio 可视化创建 Department 表(注意:这里采用的是 SQL Server 2008,如果用的是 MySQL 或 Oracle,则截取相对应的截图当作结果放在这里)

3. [实验内容3描述]

...[其他实验内容]

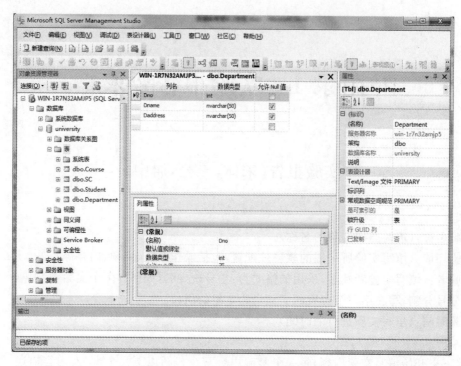

附图 1-1　采用图形界面创建数据表

附录2　课程设计报告要求

1. 题目要求

题目名称及题目要求描述。

2. 系统设计

(1)画出功能模块图。

(2)对每个功能点进行详细说明。

3. 数据库设计

(1)画出 E-R 图。

(2)列出数据库及数据表结构。

(3)采用 SQL 语句构建数据库结构(含数据库或模式、数据库用户、数据表、索引、视图、存储过程或函数、触发器等)。

4. 系统功能实现

按照系统设计,采用高级程序设计语言(C♯、C/C++、Java,三选一)设计实现各个功能模块或功能点。

5. 系统测试

(1)数据库测试数据加载(采用 SQL 语言,编写脚本插入测试数据)。

(2)系统功能模块测试。

6. 总结

第二部分

参考答案

1 MySQL 的参考答案

由于数据库管理系统软件种类众多，本书除了在每个章节给出了 Oracle 版本的答案之外，还提供了另外两种常用的数据库管理系统 MySQL 和 SQL Server 的参考答案，以方便读者使用这两种数据库管理系统进行数据库实验。

数据库管理系统软件参考答案

见第一部分 1.2

数据库和数据表操作参考答案

1. 答案

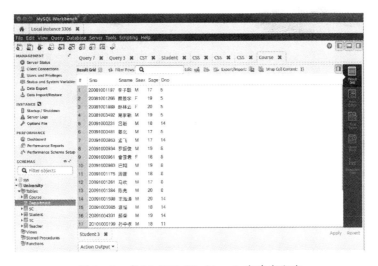

图 1-1 在 MySQL Workbench 中建库和表

2. 答案
```
use University;
drop table SC;
drop table TC;
drop table Course;
```

drop table Teacher;
drop table Student;
drop table Department;
drop database University;

3. 答案

create databaseUniversity;

4. 答案
/* 系的信息表 Department(Dno,Dname,Daddress) */
create tableDepartment(
   Dno int,
   Dname char(50),
   Daddress char(50),
   primary key(Dno)
   );
/* 学生信息表 Student(Sno,Sname,Ssex,Sage,Dno) */
create table Student(
   Sno char(11),
   Sname char(8),
   Ssex char(2),
   Sage int,
   Dno int,
   primary key(Sno),
   foreign key(Dno)references Department(Dno)
   );
/* 教师信息表 Teacher(Tno,Tname,Ttitle,Dno) */
create table Teacher(
   Tno int primary key,
   Tname char(8),
   Ttitle char(8),
   Dno int,
   foreign key(Dno)references Department(Dno)
   );
/* 课程信息表 Course (Cno,Cname,Cpno,Ccredit) */
create table Course(
   Cno int primary key,
   Cname char(50),
   Cpno int,

CCredit int,
   foreign key(Cpno) references Course(Cno)
);
/* 学生选课表 SC(Sno,Cno,Grade) */
create table SC(
   Sno char(11),
   Cno int,
   Grade int,
   primary key(Sno,Cno),
   foreign key(Sno) references Student(Sno),
   foreign key(Cno) references Course(Cno)
);
/* 教师授课表 TC(Tno,Cno,Site) */
create table TC(
   Tno int,
   Cno int,
   Site char(50),
   primary key(Tno,Cno),
   foreign key(Tno) references Teacher(Tno),
   foreign key(Cno) references Course(Cno)
);

5. 答案
create unique index stuname on Student(Sname);

6. 答案
drop index Student.stuname;

7. 答案
alter table teacher add Tsex char(2);

8. 答案
alter table teacher drop column Tsex;

数据表的数据操作参考答案

1. 答案
insert into Department(Dno,Dname,Daddress) values(1,'地球科学学院','主楼东');
insert into Department(Dno,Dname,Daddress) values(2,'资源学院','主楼西');

insert into Department(Dno,Dname,Daddress)values(3,'材化学院','材化楼');
insert into Department(Dno,Dname,Daddress)values(4,'环境学院','文华楼');
insert into Department(Dno,Dname,Daddress)values(5,'工程学院','水工楼');
insert into Department(Dno,Dname,Daddress)values(6,'地球物理与空间信息学院','物探楼');
insert into Department(Dno,Dname,Daddress)values(7,'机械与电子信息学院','教二楼');
insert into Department(Dno,Dname,Daddress)values(8,'经济管理学院','经管楼');
insert into Department(Dno,Dname,Daddress)values(9,'外语学院','北一楼');
insert into Department(Dno,Dname)values(10,'信息工程学院');
insert into Department(Dno,Dname,Daddress)values(11,'数学与物理学院','基委楼');
insert into Department(Dno,Dname,Daddress)values(12,'珠宝学院','珠宝楼');
insert into Department(Dno,Dname,Daddress)values(13,'政法学院','政法楼');
insert into Department(Dno,Dname,Daddress)values(14,'计算机学院','北一楼');
insert into Department(Dno,Dname)values(15,'远程与继续教育学院');
insert into Department(Dno,Dname)values(16,'国际教育学院');
insert into Department(Dno,Dname,Daddress)values(17,'体育部','体育馆');
insert into Department(Dno,Dname,Daddress)values(18,'艺术与传媒学院','艺传楼');
insert into Department(Dno,Dname,Daddress)values(19,'马克思主义学院','保卫楼');
insert into Department(Dno,Dname,Daddress)values(20,'江城学院','江城校区');

2. 答案

insert into Student(Sno,Sname,Ssex,Sage,Dno)values('20091000231','吕岩','M',18,14);
insert into Student(Sno,Sname,Ssex,Sage,Dno)values('20091004391','颜荣','M',19,14);
insert into Student(Sno,Sname,Ssex,Sage,Dno)values('20091001598','王海涛','M',20,14);
insert into Student(Sno,Sname,Ssex,Sage,Dno)values('20091003085','袁恒','M',18,14);
insert into Student(Sno,Sname,Ssex,Sage,Dno)values('20091000863','孟飞','M',17,14);
insert into Student(Sno,Sname,Ssex,Sage,Dno)values('20091000934','罗振俊','M',19,8);
insert into Student(Sno,Sname,Ssex,Sage,Dno)values('20091000961','曾雪君','F',18,8);
insert into Student(Sno,Sname,Ssex,Sage,Dno)values('20091000983','巴翔','M',19,8);
insert into Student(Sno,Sname,Ssex,Sage,Dno)values('20091001175','周雷','M',18,8);
insert into Student(Sno,Sname,Ssex,Sage,Dno)values('20091001261','马欢','M',17,8);
insert into Student(Sno,Sname,Ssex,Sage,Dno)values('20091001384','陈亮','M',20,8);
insert into Student(Sno,Sname,Ssex,Sage,Dno)values('20081003492','易家新','M',19,5);
insert into Student(Sno,Sname,Ssex,Sage,Dno)values('20081001197','李子聪','M',17,5);
insert into Student(Sno,Sname,Ssex,Sage,Dno)values('20081001266','蔡景学','F',19,5);
insert into Student(Sno,Sname,Ssex,Sage,Dno)values('20081001888','赵林云','F',20,5);
insert into Student(Sno,Sname,Ssex,Sage,Dno)values('20091000481','姜北','M',17,5);
insert into Student(Sno,Sname,Ssex,Sage,Dno)values('20101000199','孙中孝','M',18,11);
insert into Student(Sno,Sname,Ssex,Sage,Dno)values('20101000424','杨光','M',17,11);
insert into Student(Sno,Sname,Ssex,Sage,Dno)values('20101000481','张永强','M',16,11);

insert into Student(Sno,Sname,Ssex,Sage,Dno)values('20101000619','陈博','M',20,11);
insert into Student(Sno,Sname,Ssex,Sage,Dno)values('20101000705','汤文盼','M',18,11);
insert into Student(Sno,Sname,Ssex,Sage,Dno)values('20101000802','苏海恩','M',17,11);

3. 答案
insert into Course(Cno,Cname,Ccredit)values(2,'高等数学',8);
insert into Course(Cno,Cname,Ccredit)values(6,'C 语言程序设计',4);
insert into Course(Cno,Cname,Ccredit)values(7,'大学物理',8);
insert into Course(Cno,Cname,Ccredit)values(8,'大学化学',3);
insert into Course(Cno,Cname,Ccredit)values(10,'软件工程',2);
insert into Course(Cno,Cname,Ccredit)values(12,'美国简史',2);
insert into Course(Cno,Cname,Ccredit)values(13,'中国通史',6);
insert into Course(Cno,Cname,Ccredit)values(14,'大学语文',3);
insert into Course(Cno,Cname,Cpno,Ccredit)values(5,'数据结构',6,4);
insert into Course(Cno,Cname,Cpno,Ccredit)values(4,'操作系统',5,4);
insert into Course(Cno,Cname,Cpno,Ccredit)values(1,'数据库原理',5,4);
insert into Course(Cno,Cname,Cpno,Ccredit)values(3,'信息系统',1,2);
insert into Course(Cno,Cname,Cpno,Ccredit)values(9,'汇编语言',6,2);
insert into Course(Cno,Cname,Cpno,Ccredit)values(11,'空间数据库',1,3);

4. 答案
insert into Teacher(Tno,Tname,Ttitle,Dno)values(1,'何小峰','副教授',14);
insert into Teacher(Tno,Tname,Ttitle,Dno)values(2,'刘刚才','教授',14);
insert into Teacher(Tno,Tname,Ttitle,Dno)values(3,'李星星','教授',11);
insert into Teacher(Tno,Tname,Ttitle,Dno)values(4,'翁平正','讲师',14);
insert into Teacher(Tno,Tname,Ttitle,Dno)values(5,'李川川','讲师',14);
insert into Teacher(Tno,Tname,Ttitle,Dno)values(6,'王媛媛','讲师',14);
insert into Teacher(Tno,Tname,Ttitle,Dno)values(7,'孔夏芳','副教授',14);

5. 答案
insert into SC values('20091003085',1,90);
insert into SC values('20091000863',1,98);
insert into SC values('20091000934',1,89);
insert into SC values('20091000961',1,85);
insert into SC values('20081001197',1,79);
insert into SC values('20081001266',1,97);
insert into SC values('20081001888',1,60);
insert into SC values('20091000481',1,78);
insert into SC values('20101000199',1,65);

insert into SC values('20101000424',1,78);
insert into SC values('20101000481',1,69);
insert into SC values('20091000863',6,90);
insert into SC values('20091000934',6,90);
insert into SC values('20091000961',6,87);

6. 答案
insert into TC values(1,1,'教一楼')
insert into TC values(1,6,'教一楼');
insert into TC values(2,10,'教二楼');
insert into TC values(3,2,'教三楼');
insert into TC values(4,5,'教三楼');
insert into TC values(6,3,'综合楼');
insert into TC values(7,4,'教二楼');
insert into TC values(5,9,'教一楼');

7. 答案
select * from student;

8. 答案
select student. sname from student where student. ssex='F';

9. 答案
select Dno,count(Sno)from student group by Dno;

10. 答案
select Dno,count(Tno)from Teacher group by Dno;

11. 答案
select Student. Sname,SC. Grade
from   Student,SC
where (SC. Grade>=60 and SC. Grade<=100)
and(SC. Cno=1)and(SC. Sno=Student. Sno)
order by SC. Grade DESC;

12. 答案
采用 SQL 语言编写一个连接查询:查询经济管理学院年龄在 20 岁以下的男生的姓名和年龄。

select Student.Sname,Student.Sage
from Student,Department
where
  Student.Sage<=20 and
  Student.Dno=Department.Dno and
  Department.Dname='经济管理学院';

13. 答案

采用 SQL 语言编写一个嵌套查询:查询选修课程总学分在五个学分以上的学生的姓名。

select Sname
from Student
where Sno in(
  select Sno from SC,Course where
    SC.Cno=Course.Cno
  group by Sno
  having SUM(CCredit)>=5);

14. 答案

采用 SQL 语言编写一个嵌套查询:查询各门课程的最高成绩的学生姓名及其成绩。

select Cno,Sname,Grade
from Student,SC SCX
where Student.Sno=SCX.Sno and SCX.Grade in
(
 select MAX(Grade)
 from SC SCY
 where SCX.Cno=SCY.Cno
 group by Cno
);

15. 答案

采用 SQL 语言查询所有选修了何小峰老师开设课程的学生姓名及其所在的院系名称。
select Sname,Dname
from Student,Department,SC
where Student.Sno=SC.Sno and Student.Dno=Department.Dno
and SC.Cno in(
select Cno
from TC

where TC. Tno=(select Tno from Teacher where Tname='何小峰')
);

16. 答案
采用SQL语言,在数据库中删除学号为20091003085的学生的所有信息(包括其选课记录)。

delete from SC where SC. Sno='20091003085';
delete from Student where Sno='20091003085';

17. 答案
采用SQL语言,将学号为20091000863的学生的"数据库原理"这门课的成绩修改为80分。

update SC set Grade=80
where Sno='20091000863'
and Cno=(select Cno from Course where Cname='数据库原理');

视图的创建与使用参考答案

1. 答案

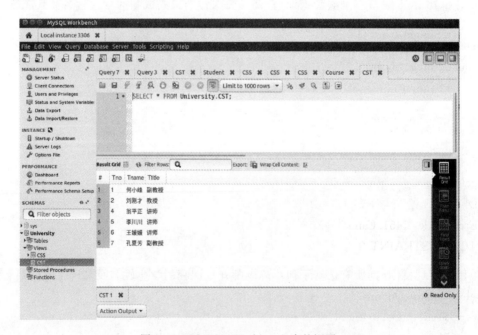

图 1-2　MySQL Workbench 中的视图

2. 答案
create view CSS as

```
select Sno,Sname,Ssex,Sage
from Student
where Dno='14';
(
select Dno from Department where Dname='计算机学院'
);
insert into CSS values('20101000911','钟晓年','M',16);
delete from CSS where Sno='20101000911';
drop view CSS;
```

3. 答案
```
create view CST as
select Tno,Tname,Ttitle
from Teacher
where Dno=
(
select Dno from Department where Dname='计算机学院'
);
```

4. 答案
```
select Sname from CSS where Sno in
(select Sno from SC SCX where SCX.Cno in(
select distinct SCY.Cno from SC SCY,TC
where SCY.Cno=TC.Cno and TC.Tno in
(select Tno from CST)));
```

5. 答案
```
delete from CST where Tno=1;
```
DELETE 语句与 REFERENCE 约束"FK__TC__Tno__1920BF5C"冲突。

Error Code: 1451. Cannot delete or update a parent row: a foreign key constraint fails('University'.'TC',CONSTRAINT 'TC_ibfk_1' FOREIGN KEY('Tno')REFERENCES 'Teacher'('Tno'))

数据库安全性参考答案

1. 答案
```
create user U1 identified by 'cug';
create user U2 identified by 'cug';
create user U3 identified by 'cug';
```

```
create user U4 identified by 'cug';
create user U5 identified by 'cug';
create user U6 identified by 'cug';
create user U7 identified by 'cug';
```

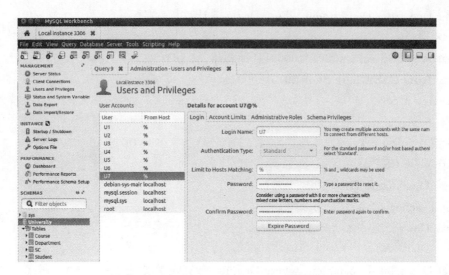

图 1-3　MySQL Workbench 中创建用户

MySQL 没有赋予 connect 这个权限,一般有 select、insert、update、delete、create、drop 等权限。

2. 答案
```
grant select on Student to U1;
```

3. 答案
```
grant all on Student to U2,U3;
grant all on Course to U2,U3;
```

4. 答案
MySQL 不支持直接将权限授予所有用户,只能每次授予一个,然后重复执行。

```
grant select on SC to U1;
grant select on SC to U2;
grant select on SC to U3;
grant select on SC to U4;
grant select on SC to U5;
grant select on SC to U6;
grant select on SC to U7;
```

5. 答案

grant select,update on SC to U4;

6. 答案

grant insert on SC to U5 with grant option;

7. 答案

grant insert on SC to U6;

8. 答案

U6 不能对 U7 进行转授权。

9. 答案

revoke update on SC from U4;

revoke select on SC from U1;
revoke select on SC from U2;
revoke select on SC from U3;
revoke select on SC from U4;
revoke select on SC from U5;
revoke select on SC from U6;
revoke select on SC from U7;

11. 答案
略

数据库完整性参考答案

1. 答案
（a）
create table STU_T(
　　Sno int primary key,
　　Sname varchar(50),
　　Ssex varchar(2),
　　Sage int,
　　Dno int
);
drop table STU_T;

(b)
```
create table STU_T (
    Sno int,
    Sname varchar(50),
    Ssex varchar(2),
    Sage int,
    Dno int,
    primary key(Sno)
);
drop table STU_T;
```
(c)
```
create table STU_T (
    Sno int,
    Sname varchar(50),
    Ssex varchar(2),
    Sage int,
    Dno int
);
alter table STU_T add constraint pk_sno primary key(Sno);
```
(d)
不支持删除 CONSTRAINT,只能删 foreign key constraint。
```
drop table STU_T;
```

2. 答案
(a)
```
create table SC_T(
Sno int,
Cno int,
Grade float,
primary key(Sno,Cno));
drop table SC_T;
```
(b)
```
create table SC_T(
    Sno int,
    Cno int,
    Grade float);
alter table SC_T add constraint pk_sc primary key(Sno,Cno);
drop table SC_T;
```

3. 答案

(a)

create table TC_T(TNO INT,CNO INT,SITE VARCHAR(50),
    primary key(TNO,CNO),
    foreign key(TNO)references　Teacher(Tno),
    foreign key(CNO)references　Course(Cno));

drop table TC_T;

(b)

create table TC_T(Tno lnt,Cno int,Site varchar(50));
alter table TC_T add constraint pk_tc primary key(Tno,Cno);
alter table TC_T add constraint fk_tno foreign key(tno)references Teacher(Tno);
alter table TC_T add constraint fk_cno foreign key(Cno)references Course(Cno);
drop table TC_T;

4. 答案

(a)

create table DEP_T(
    Dno int primary key,
    Dname varchar(50)unique not null,
    Daddress varchar(50)default '北一楼');

drop table　DEP_T cascade;

(b)

create table DEP_T(
    DNO int,
    DNAME varchar(50),
    DADDRESS varchar(50),
    primary key(Dno),
    check(Dno>0 and Dno<100));

drop table　DEP_T cascade;

触发器参考答案

1. 答案

create table SGA_T(Sno char(11)primary key,AverageGrade float);
insert into SGA_T select Sno,avg(Grade)from SC group by Sno;

(a)

delimiter $
create trigger SC_UPDATE_TRIGGER after update

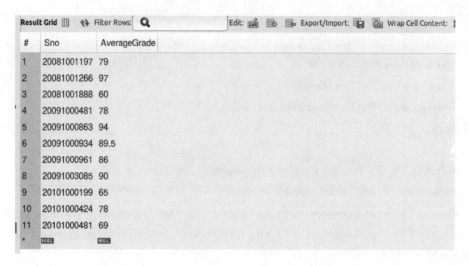

图 1-4 MySQL Workbench 中触发器执行结果

on SC for each row
begin
delete from SGA_T;
insert into SGA_T select Sno,avg(Grade)from SC group by Sno;
end$
delimiter;
(b)
delimiter $
create trigger SC_INSERT_TRIGGER after insert
on SC for each row
begin
delete from SGA_T;
insert into SGA_T select Sno,avg(Grade)from SC group by Sno;
end$
delimiter;
(c)
delimiter $
create trigger SC_DELETE_TRIGGER after delete
on SC for each row
begin
delete from SGA_T;
insert into SGA_T select Sno,avg(Grade)from SC group by Sno;
end$
delimiter;

(d)
drop trigger SC_INSERT_TRIGGER;
drop trigger SC_UPDATE_TRIGGER;
drop trigger SC_DELETE_TRIGGER;
drop table SGA_T;

2. 答案
(a)
delimiter $
create trigger TC_INSERT_TRIGGER after insert
on TC for each row
begin
declare CC int;
select count(*)into CC from TC where TC.Tno=NEW.Tno;
if CC>=3 then
SIGNAL SQLSTATE '02000' set message_text='NO ACTION';
end if;
end $
delimiter;
(b)
delimiter $
create trigger TC_DELETE_TRIGGER before delete
on TC for each row
begin
declare CC int;
select count(*)into CC from TC where TC.Tno=OLD.Tno;
if CC<=1 then
SIGNAL SQLSTATE '02000' set message_text='NO ACTION';
end if;
end $
delimiter;
(c)
drop trigger TC_INSERT_TRIGGER;
drop trigger TC_DELETE_TRIGGER;

3. 答案
delimiter $
create trigger TEACHER_TRIGGER after insert
on Teacher for each row

begin
declare CC int;
select count(*)into CC from Department where Dno=NEW. Dno;
if CC<=0 then
SIGNAL SQLSTATE '02000' set message_text='NO ACTION';
end if;
end $
delimiter;

drop trigger TEACHER_TRIGGER;

数据库设计参考答案

1.答案

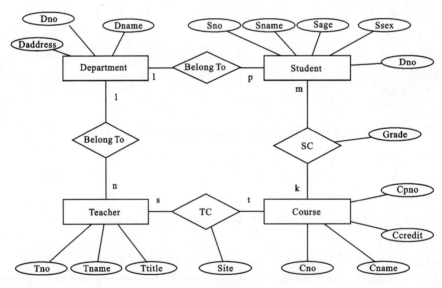

图1-5 示例数据库 E-R 图

2.答案
第一步,写出 UNIVERSITY 关系数据库模式。
(g)系的信息表 Department(Dno,Dname,Daddress);
(h)学生信息表 Student(Sno,Sname,Ssex,Sage,Dno);
(i)教师信息表 Teacher(Tno,Tname,Ttitle,Dno);
(j)课程信息表 Course(Cno,Cname,Cpno,Ccredit);
(k)学生选课表 SC(Sno,Cno,Grade);
(l)教师授课表 TC(Tno,Cno,Site)。

第二步,结合选定的数据库管理系统(Oracle),列出每个关系中每个属性的类型、长度等信息。

表 1-1 Department 关系属性表

关系名称		Department	关系别名				系的信息	
属性名	别名	类型	长度	值域	唯一	可空	备注	
Dno	系编号	INT			Y	N		
Dname	系名称	VARCHAR	50		Y	N		
Daddress	系地址	VARCHAR	50		Y	Y		

表 1-2 Student 关系属性表

关系名称		Student	关系别名				学生信息	
属性名	别名	类型	长度	值域	唯一	可空	备注	
Sno	学号	INT			Y	N		
Sname	姓名	VARCHAR	50		N	N		
Ssex	性别	VARCHAR	2	M/F	N	Y		
Sage	年龄	INT			N	Y		
Dno	系编号	INT			Y	Y	所在系	

表 1-3 Teacher 关系属性表

关系名称		Teacher	关系别名				教师信息	
属性名	别名	类型	长度	值域	唯一	可空	备注	
Tno	工号	INT			Y	N		
Tname	姓名	VARCHAR	50		N	N		
Ttitle	性别	VARCHAR	50		N	Y		
Dno	系编号	INT			Y	Y	所在系	

表 1-4 Course 关系属性表

关系名称		Course	关系别名				课程信息	
属性名	别名	类型	长度	值域	唯一	可空	备注	
Cno	课程号	INT			Y	N		
Cname	课程名	VARCHAR	50		Y	N		
Cpno	先导课	INT			N	Y		
Ccredit	学分	INT			Y	Y	所在系	

表 1-5 SC关系属性表

关系名称		Course		关系别名		课程信息	
属性名	别名	类型	长度	值域	唯一	可空	备注
Cno	课程号	INT			Y	N	
Sno	学号	INT			Y	N	
Grade	成绩	FLOAT			N	Y	

表 1-6 TC关系属性表

关系名称		Course		关系别名		课程信息	
属性名	别名	类型	长度	值域	唯一	可空	备注
Cno	课程号	INT			Y	N	
Tno	工号	INT			Y	N	
Site	位置	VARCHAR	50		N	Y	

第三步,画出数据库模式图,如图 1-6 所示。

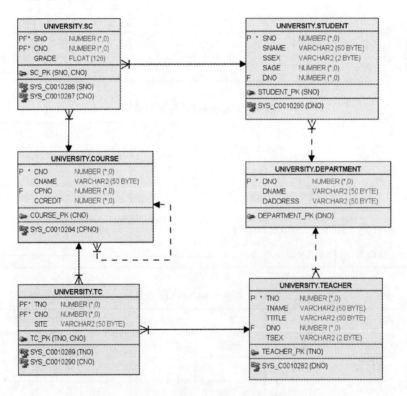

图 1-6 UNIVERSITY 关系数据库结构

第四步,设计用户子模式。例如,设计计算机系学生信息视图 CSS。这里没有需要设计的其他用户子模式,故可以直接进入下一步物理设计。

3. 答案
```
create table Department(
    Dno int,
    Dname varchar(50),
    Daddress varchar(50),
    primary key(Dno)
);
create table Student(
    Sno int,
    Sname varchar(50),
    Ssex varchar(2),
    Sage int,
    Dno int,
    primary key(Sno),
    foreign key(Dno)references Department(Dno)
);
create table Teacher(
    Tno int primary key,
    Tname varchar(50),
    Ttitle varchar(50),
    Dno int,
    foreign key(Dno)references Department(Dno)
);
create table Course(
    Cno int primary key,
    Cname varchar(50),
    Cpno int,
    CCredit int,
    foreign key(Cpno)references Course(Cno)
);
create table SC(
    Sno int,
    Cno int,
    Grade float,
    primary key(Sno,Cno),
    foreign key(Sno)references Student(Sno),
    foreign key(Cno)references Course(Cno)
);
create table TC(
```

```
    Tno int,
    Cno int,
    Site varchar(50),
    primary key(Tno,Cno),
    foreign key(Tno)references Teacher(Tno),
    foreign key(Cno)references Course(Cno)
);
```

存储过程与函数参考答案

1. 答案

```
delimiter $
create  procedure DecreaseGrade()
begin
update SC set Grade=Grade * 0.95;
end $
delimiter;

call DecreaseGrade;
```

2. 答案

```
delimiter $
create  procedure IncreaseGrade( in ccno int)
begin
update SC set Grade=Grade * 1.05 WHERE Cno=ccno;
end $
delimiter;

call IncreaseGrade(1);
```

3. 答案

```
delimiter $
create  procedure AverageStudentGrade( in paramsno varchar(20),out paramgrade float)
begin
declare g float default 0.0;
select sg.ag into g
        from(select Sno s,avg(Grade)ag
```

                from SC group by Sno)sg
            where sg.s=paramsno;
end $
delimiter;
call AverageStudentGrade(20091000863,@g);

4. 答案
drop procedure DecreaseGrade;
drop procedure IncreaseGrade;

5. 答案
定义一个带有输入参数的自定义函数 CalculateAverageStudentGrade,计算一个学生的所有选修课程的平均成绩,要求以学号作为输入参数,返回该生的所有选修课平均成绩,调用函数,并输出计算结果。

delimiter $
create function CalculateAverageStudentGrade(paramsno varchar(11))
returns int
begin
    declare g int;
    select sg.ag into g from
    (select Sno s,avg(Grade)ag from
    SC group by Sno)sg
    where sg.s=paramsno;
    return g;
end $

delimiter;

/
select   CalculateAverageStudentGrade('20091000863');

6. 答案
删除函数 CalculateAverageStudentGrade。

drop function CalculateAverageStudentGrade;

7. 答案
定义一个存储过程,采用普通无参游标实现计算学校开设的所有课程的学分之和。

```
delimiter $
create procedure pro_7()
begin
--创建接受游标数据的变量,即每次从表中读取的credit值
declare credit int;
--创建总数变量
declare credit_sum int default false;
--创建结束标志变量
declare done int default false;
--创建游标
declare cur cursor for select CCredit from Course;
--指定游标循环结束时的返回值
declare continue HANDLER for not found set done=true;
--打开游标
open cur;
--开始循环游标里的数据
read_loop:loop
fetch cur into credit;
if done then
    leave read_loop;
end if;
--获取数据以后进行累加操作
set credit_sum=credit_sum + credit;
--结束游标循环
end loop;
--关闭游标
close cur;
--返回结果
select credit_sum;
end $

delimiter;

/
call pro_7();
```

8. 答案

定义一个存储过程,采用REF CURSOR实现计算学院所有学生选修课程的成绩之和。

```
delimiter $
create procedure pro_8()
begin
declare grade int;
declare grade_sum int default 0;
declare done int default false;
declare cur_grade cursor for select Grade from SC;
declare continue HANDLER for not found set done=true;
open cur_grade;
read_loop:loop
fetch cur_grade into grade;
if done then
    leave read_loop;
end if;
set grade_sum=grade_sum + grade;
end loop;
close cur_grade;
select grade_sum;
end $
delimiter;
/
call pro_8();
```

9. 答案
定义一个存储过程,采用带参数游标实现按照学号计算学生的平均成绩。

```
delimiter $
create procedure pro_9(in sno_in char(11))
begin
declare avg_grade int default 0;
declare done boolean default false;
declare cur cursor for select sg.ag
        from(select Sno s,avg(Grade)ag
                from SC group by Sno)sg
                    where sg.s=sno_in;
declare continue HANDLER for not found set done=true;
open cur;
read_loop:loop
fetch cur into avg_grade;
```

```
if done then
    leave read_loop;
end if;
select avg_grade;
end loop;
close cur;
end $

delimiter;
/
call pro_9();
```

数据库应用开发(C++)参考答案

1. 答案

本章采用 Visual Studio 集成开发环境。下载 Connector/C++1.1.11、boost1.67.0,并且解压到制定目录。主要步骤如下:

(1)新建 TestMySQLAPP 控制台项目,或者直接新建一个空项目(图 1-7)。

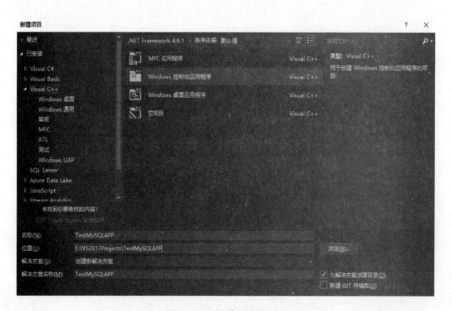

图 1-7 新建示例工程

(2)将默认的 debug 模式改为 release 模式(图 1-8)。
(3)添加 include 以及 boost 的解压目录(图 1-9、图 1-10)。
(4)添加 lib 目录(图 1-11)。
(5)添加 mysqlcppconn.lib(图 1-12)。

1 MySQL 的参考答案

图 1-8 修改模式

图 1-9 添加 MySQL Connector/C++头文件

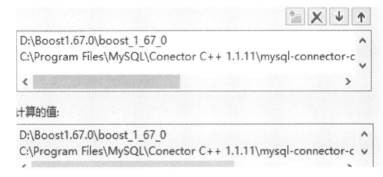

图 1-10 添加 boost 头文件

图 1-11 设置附加库目录

图 1-12 添加依赖项

(6)将 lib 目录下的所有.dll 文件拷贝粘贴到 C:\Windows\System32 目录。
(7)测试(图 1-13)。

```cpp
#include <stdafx.h>
#include <stdlib.h>
#include <iostream>

#include "mysql_connection.h"

#include <cppconn/driver.h>
#include <cppconn/exception.h>
#include <cppconn/resultset.h>
#include <cppconn/statement.h>

using namespace std;

int main()
{
    cout << endl;
    cout << "正在执行 SELECT sage from student where sno='20081001197'" << endl;

    try {
        sql::Driver  *driver;
        sql::Connection  *con;
        sql::Statement  *stmt;
        sql::ResultSet  *res;

        /*创建连接*/
        driver=get_driver_instance();
        con=driver->connect("tcp://127.0.0.1:3306","root","root");
        /*连接 MySQL 数据库 university */
        con->setSchema("university");

        stmt=con->createStatement();
        res=stmt->executeQuery("SELECT sage from student where sno='20081001197'");//标准 sql 语句
        while(res->next()){
            cout << "\t MySQL replies: ";
            /*通过数字偏移量,1 代表第一列*/
```

```
                cout << res->getInt(1)<< endl;
            }
            delete res;
            delete stmt;
            delete con;
        }
        catch(sql::SQLException &e){
            cout << "# ERR: SQLException in " << __FILE__;
            cout << "(" << __FUNCTION__ << ")on line " << __LINE__ << endl;
            cout << "# ERR: " << e.what();
            cout << "(MySQL error code: " << e.getErrorCode();
            cout << ",SQLState: " << e.getSQLState()<< " )" << endl;
        }

        cout << endl;
        return EXIT_SUCCESS;
    }
```

图 1-13 运行结果

测试成功。其他答案的思路和第 10 章一样，只需修改对应的 SQL 语句即可。

数据库应用开发(Java)参考答案

1. 答案

(1)基于 IDEA 构建 Java 工程。在 IDEA 中新建工程，选择 Java 类型，设置 Project SDK (图 1-14);然后选择工程模版(图 1-15);接着，输入工程名称和存放位置(图 1-16);最后生成如图 1-17 所示的 Hello World 工程。

(2)点击"File→Project Structure→Libraries"，再点击加号，添加 jdbc8.jar 和其他相关的依赖 Jar 包(图 1-18、图 1-19)。

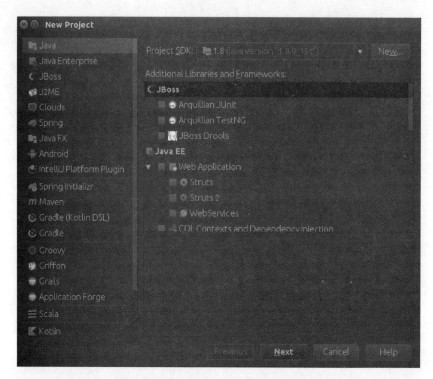

图 1-14　IDEA 构建 Java 工程

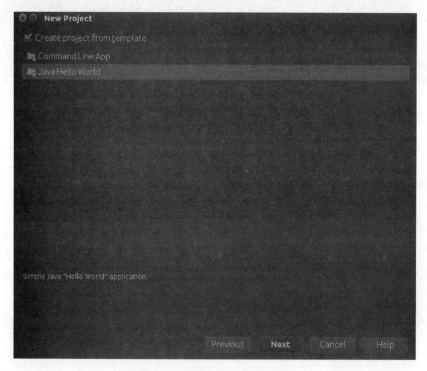

图 1-15　IDEA Java 工程模板

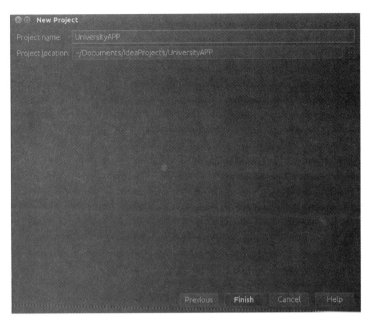

图 1-16　IDEA Java 工程基本信息

图 1-17　IDEA Java HelloWorld

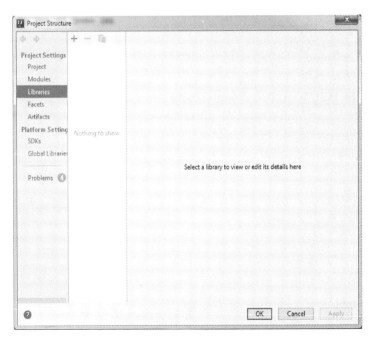

图 1-18　IDEA Java 工程添加依赖包

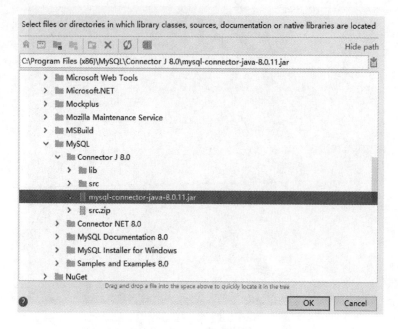

图 1-19　IDEA Java 工程添加 JDBC 下的依赖包

(3)连接本书示例数据库 UNIVERSITY 的代码。

```
public static Connection getConnection()throws Exception{
    Connection conn=null; //创建一个 connection
    try {
        Class.forName("com.mysql.cj.jdbc.Driver");
        String url="jdbc:mysql://localhost:3306/University?";
        String userName="Your-User-Name";
        String password  ="Your-Password";
        conn=DriverManager.getConnection(url,userName,password);
    } catch(ClassNotFoundException e){
        System.out.println("JDBC Driver not found");
        e.printStackTrace();
    } catch(SQLException e){
        e.printStackTrace();
    } catch(Exception e){
        e.printStackTrace();
    }
    return conn;
}
```

构建语句,执行并获取查询结果:

```
public static void main(String[] args){
    String sql="select sno,sum(grade)sg from sc group by sno";
    Connection connection=getConnection();
    try {
        Statement statement=connection.createStatement();
        ResultSet resultSet=    statement.executeQuery(sql);
        while(resultSet.next()){
            String sno=resultSet.getString("sno");
            double sg=resultSet.getDouble("sg");
            System.out.println(sno);
            System.out.println(sg);
        }
    }
    catch(SQLException e){
        e.printStackTrace();
    }
}
```

2. 答案

解决本题的思路和方法与第 1 题是一样的,唯一不同的地方是更换 SQL 语句。将上一题的 SQL 语句替换成如下语句:

select student.sname,sa.ag from student,(select sno,avg(grade)from sc group by sno)as sa(sno,ag)where sa.sno=student.sno;

数据库应用开发(C♯)参考答案

1. 答案

需要安装 Connector/NET 8.0.11,可以在安装 MySQL 时就选中,如果没有选中也可以到官网下载安装:https://dev.mysql.com/downloads/connector/net/。

第一步,构建 C♯ 程序,选择 Windows Form App(.NET Framework),如图 1-20 所示。构建 UniApp 工程,然后将"Form"改为"MainForm",如图 1-21 所示。

第二步,添加引用,如图 1-22 所示,选择"Project→Add Reference..."弹出如图 1-23 所示的界面,选择"MySql.Data"。

图 1-20 选择"Windows Form App(.NET Framework)"工程类型

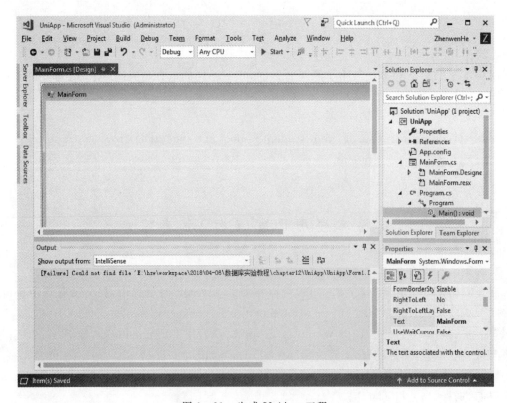

图 1-21 生成 UniApp 工程

1　MySQL 的参考答案

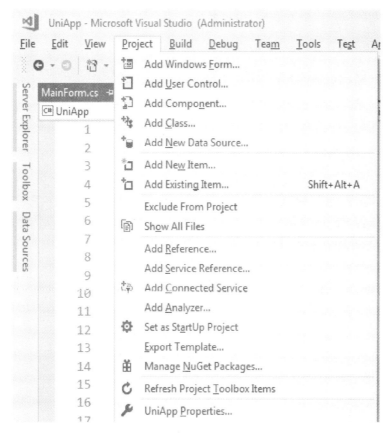

图 1-22　Add Reference 菜单

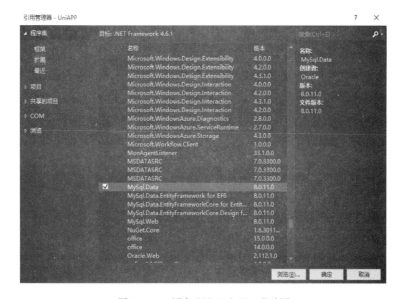

图 1-23　添加"MySql.Data"引用

第三步,在 MainFrame.cs 中添加获取连接的函数,该函数通过传入数据库服务器所在主机名或 IP 地址、监听端口、服务名称、用户名称和用户密码,构建一个 Connection 对象并打开。具体编码如下:

```
public static MySqlConnection GetConnection(
    string host,string port,string serviceName,
    string user,string password)
{
    string constr=
    "Data Source=(DESCRIPTION=(ADDRESS=(PROTOCOL=TCP)(HOST="
    + host+" )(PORT="
    + port+" ))(CONNECT_DATA =(SERVER=DEDICATED)(SERVICE_NAME="
    + serviceName +" ))); "
    +"User Id="
    + user+"; Password="
    +password+";";
    try
    {
        MySqlConnection conn=new MySqlConnection(constr);
        conn. ConnectionString=constr;
        conn. Open();
        return conn;
    }
    catch(Exception ex)
    {
        Console. WriteLine(ex. Message);
        Console. WriteLine(ex. Source);
    }
    return null;
}
```

第四步,在 MainFrame.cs 中添加获取查询函数。该函数在获取数据库连接后,构建一个 Command 对象,执行获取学生平均成绩的查询语句,返回一个 DataReader 对象,最后获取返回结果。具体编码如下:

```
public static void ExecuteQuery(string sql)
{
    if(sql==null)
        sql="select sno,avg(grade)sg from sc group by sno;";
```

```
        try
        {
            MySqlConnection conn=GetConnection("localhost",
                            "1521","pdborcl","UNIVERSITY","cug");
            MySqlCommand comm=new MySqlCommand(sql,conn);
            MySqlDataReader dr=comm. ExecuteReader();
            while(dr. Read())
            {
                long  d1=dr. GetInt64(0);
                float d2=dr. GetFloat(1);
                Console. WriteLine(d1);
                Console. WriteLine(d2);
            }
        }
        catch(Exception ex)
        {
            Console. WriteLine(ex. Message);
            Console. WriteLine(ex. Source);
        }
    }
```

数据备份与恢复参考答案

1. 答案

格式：mysqldump-h 主机名-P 端口-u 用户名-p 密码-database 数据库名 > 文件名.sql。主机以及端口名可以省略不写：

mysqldump-u root-p University>University. sql

接下来会提示输入数据库 root 密码,最后会将数据库备份到当前目录名为 University.sql 的文件中,或者可以自定义路径。

2. 答案
mysqldump-u root-p University Student Course>sc. sql

3. 答案
还原使用 mysqldump 命令备份的数据库的语法如下：

mysql - u root - p [dbname] < backup.sql

另外可以使用 source 命令来导入数据库。常用 source 命令 mysql>source d:\test.sql，后面的参数为脚本文件。

source ~/Documents/University.sql

4. 答案
source~/Documents/SC.sql

综合实验(课程设计)参考答案(略)

2 SQL Server 的参考答案

数据库管理系统软件参考答案

数据库管理系统软件参考答案(略)

数据库和数据表操作参考答案

1. 答案

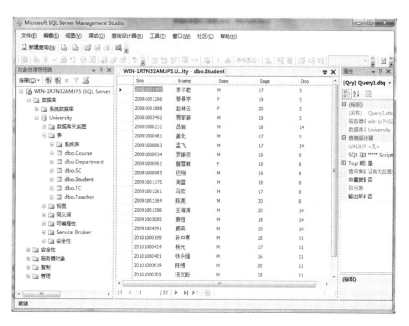

图 2-1 在 Microsoft SQL Server Management Studio 中建库和建表

2. 答案
use UNIVERSITY;
drop table SC;
drop table TC;
drop table Course;

```sql
drop table Teacher;
drop table Student;
drop table Department;
drop database UNIVERSITY;
```

3. 答案
```sql
create database UNIVERSITY;
```

4. 答案
```sql
/* 系的信息表 Department(Dno,Dname,Daddress) */
create table Department(
    Dno int,
    Dname char(50),
    Daddress char(50),
    primary key(Dno)
);
/* 学生信息表 Student(Sno,Sname,Ssex,Sage,Dno) */
create table Student(
    Sno char(11),
    Sname char(8),
    Ssex char(2),
    Sage int,
    Dno int,
    primary key(Sno),
    foreign key(Dno) references Department(Dno)
);
/* 教师信息表 Teacher(Tno,Tname,Ttitle,Dno) */
create table Teacher(
    Tno int primary key,
    Tname char(8),
    Ttitle char(8),
    Dno int,
    foreign key(Dno) references Department(Dno)
);
/* 课程信息表 Course (Cno,Cname,Cpno,Ccredit) */
create table Course(
    Cno int primary key,
    Cname char(50),
    Cpno int,
```

CCredit int,
　　foreign key(Cpno) references Course(Cno)
　　);
/* 学生选课表 SC(Sno,Cno,Grade) */
create table SC(
　　Sno char(11),
　　Cno int,
　　Grade int,
　　primary key(Sno,Cno),
　　foreign key(Sno) references Student(Sno),
　　foreign key(Cno) references Course(Cno)
　　);
/* 教师授课表 TC(Tno,Cno,Site) */
create table TC(
　　Tno int,
　　Cno int,
　　Site char(50),
　　primary key(Tno,Cno),
　　foreign key(Tno) references Teacher(Tno),
　　foreign key(Cno) references Course(Cno)
　　);

5. 答案
create unique index Stuname on Student(Sname);

6. 答案
drop index Student.Stuname;

7. 答案
alter table Teacher add Tsex char(2);

8. 答案
alter table Teacher drop column Tsex;

数据表的数据操作参考答案

1. 答案
insert into Department(Dno,Dname,Daddress) values(1,'地球科学学院','主楼东');

insert into Department(Dno,Dname,Daddress)values(2,'资源学院','主楼西');
insert into Department(Dno,Dname,Daddress)values(3,'材化学院','材化楼');
insert into Department(Dno,Dname,Daddress)values(4,'环境学院','文华楼');
insert into Department(Dno,Dname,Daddress)values(5,'工程学院','水工楼');
insert into Department(Dno,Dname,Daddress)values(6,'地球物理与空间信息学院','物探楼');
insert into Department(Dno,Dname,Daddress)values(7,'机械与电子信息学院','教二楼');
insert into Department(Dno,Dname,Daddress)values(8,'经济管理学院','经管楼');
insert into Department(Dno,Dname,Daddress)values(9,'外语学院','北一楼');
insert into Department(Dno,Dname)values(10,'信息工程学院');
insert into Department(Dno,Dname,Daddress)values(11,'数学与物理学院','基委楼');
insert into Department(Dno,Dname,Daddress)values(12,'珠宝学院','珠宝楼');
insert into Department(Dno,Dname,Daddress)values(13,'政法学院','政法楼');
insert into Department(Dno,Dname,Daddress)values(14,'计算机学院','北一楼');
insert into Department(Dno,Dname)values(15,'远程与继续教育学院');
insert into Department(Dno,Dname)values(16,'国际教育学院');
insert into Department(Dno,Dname,Daddress)values(17,'体育部','体育馆');
insert into Department(Dno,Dname,Daddress)values(18,'艺术与传媒学院','艺传楼');
insert into Department(Dno,Dname,Daddress)values(19,'马克思主义学院','保卫楼');
insert into Department(Dno,Dname,Daddress)values(20,'江城学院','江城校区');

2. 答案

insert into Student(Sno,Sname,Ssex,Sage,Dno)values('20091000231','吕岩','M',18,14);
insert into Student(Sno,Sname,Ssex,Sage,Dno)values('20091004391','颜荣','M',19,14);
insert into Student(Sno,Sname,Ssex,Sage,Dno)values('20091001598','王海涛','M',20,14);
insert into Student(Sno,Sname,Ssex,Sage,Dno)values('20091003085','袁恒','M',18,14);
insert into Student(Sno,Sname,Ssex,Sage,Dno)values('20091000863','孟飞','M',17,14);
insert into Student(Sno,Sname,Ssex,Sage,Dno)values('20091000934','罗振俊','M',19,8);
insert into Student(Sno,Sname,Ssex,Sage,Dno)values('20091000961','曾雪君','F',18,8);
insert into Student(Sno,Sname,Ssex,Sage,Dno)values('20091000983','巴翔','M',19,8);
insert into Student(Sno,Sname,Ssex,Sage,Dno)values('20091001175','周雷','M',18,8);
insert into Student(Sno,Sname,Ssex,Sage,Dno)values('20091001261','马欢','M',17,8);
insert into Student(Sno,Sname,Ssex,Sage,Dno)values('20091001384','陈亮','M',20,8);
insert into Student(Sno,Sname,Ssex,Sage,Dno)values('20081003492','易家新','M',19,5);
insert into Student(Sno,Sname,Ssex,Sage,Dno)values('20081001197','李子聪','M',17,5);
insert into Student(Sno,Sname,Ssex,Sage,Dno)values('20081001266','蔡景学','F',19,5);
insert into Student(Sno,Sname,Ssex,Sage,Dno)values('20081001888','赵林云','F',20,5);
insert into Student(Sno,Sname,Ssex,Sage,Dno)values('20091000481','姜北','M',17,5);
insert into Student(Sno,Sname,Ssex,Sage,Dno)values('20101000199','孙中孝','M',18,11);
insert into Student(Sno,Sname,Ssex,Sage,Dno)values('20101000424','杨光','M',17,11);

insert into Student(Sno,Sname,Ssex,Sage,Dno)values('20101000481','张永强','M',16,11);
insert into Student(Sno,Sname,Ssex,Sage,Dno)values('20101000619','陈博','M',20,11);
insert into Student(Sno,Sname,Ssex,Sage,Dno)values('20101000705','汤文盼','M',18,11);
insert into Student(Sno,Sname,Ssex,Sage,Dno)values('20101000802','苏海恩','M',17,11);

3. 答案
insert into Course(Cno,Cname,Ccredit)values(2,'高等数学',8);
insert into Course(Cno,Cname,Ccredit)values(6,'C 语言程序设计',4);
insert into Course(Cno,Cname,Ccredit)values(7,'大学物理',8);
insert into Course(Cno,Cname,Ccredit)values(8,'大学化学',3);
insert into Course(Cno,Cname,Ccredit)values(10,'软件工程',2);
insert into Course(Cno,Cname,Ccredit)values(12,'美国简史',2);
insert into Course(Cno,Cname,Ccredit)values(13,'中国通史',6);
insert into Course(Cno,Cname,Ccredit)values(14,'大学语文',3);
inscrt into Course(Cno,Cname,Cpno,Ccredit)values(5,'数据结构',6,4);
insert into Course(Cno,Cname,Cpno,Ccredit)values(4,'操作系统',5,4);
insert into Course(Cno,Cname,Cpno,Ccredit)values(1,'数据库原理',5,4);
insert into Course(Cno,Cname,Cpno,Ccredit)values(3,'信息系统',1,2);
insert into Course(Cno,Cname,Cpno,Ccredit)values(9,'汇编语言',6,2);
insert into Course(Cno,Cname,Cpno,Ccredit)values(11,'空间数据库',1,3);

4. 答案
insert into Teacher(Tno,Tname,Ttitle,Dno)values(1,'何小峰','副教授',14);
insert into Teacher(Tno,Tname,Ttitle,Dno)values(2,'刘刚才','教授',14);
insert into Teacher(Tno,Tname,Ttitle,Dno)values(3,'李星星','教授',11);
insert into Teacher(Tno,Tname,Ttitle,Dno)values(4,'翁平正','讲师',14);
insert into Teacher(Tno,Tname,Ttitle,Dno)values(5,'李川川','讲师',14);
insert into Teacher(Tno,Tname,Ttitle,Dno)values(6,'王媛媛','讲师',14);
insert into Teacher(Tno,Tname,Ttitle,Dno)values(7,'孔夏芳','副教授',14);

5. 答案
insert into SC values('20091003085',1,90);
insert into SC values('20091000863',1,98);
insert into SC values('20091000934',1,89);
insert into SC values('20091000961',1,85);
insert into SC values('20081001197',1,79);
insert into SC values('20081001266',1,97);
insert into SC values('20081001888',1,60);
insert into SC values('20091000481',1,78);

insert into SC values('20101000199',1,65);
insert into SC values('20101000424',1,78);
insert into SC values('20101000481',1,69);
insert into SC values('20091000863',6,90);
insert into SC values('20091000934',6,90);
insert into SC values('20091000961',6,87);

6. 答案
insert into TC values(1,1,'教一楼')
insert into TC values(1,6,'教一楼');
insert into TC values(2,10,'教二楼');
insert into TC values(3,2,'教三楼');
insert into TC values(4,5,'教三楼');
insert into TC values(6,3,'综合楼');
insert into TC values(7,4,'教二楼');
insert into TC values(5,9,'教一楼');

7. 答案
select * from Student;

8. 答案
select Student.Sname from Student where Student.Ssex='F';

9. 答案
select Dno,count(Sno)from Student group by Dno;

10. 答案
select Dno,count(Tno)from Teacher group by Dno;

11. 答案
select Student.Sname,SC.Grade
from Student,SC
where(SC.Grade>=60 and SC.Grade<=100)
and(SC.Cno=1)and(SC.Sno=Student.Sno)
order by SC.Grade desc;

12. 答案
采用SQL语言编写一个连接查询:查询经济管理学院年龄在20岁以下的男生的姓名和年龄。

```
select Student.Sname,Student.Sage
from Student,Department
where
    Student.Sage<=20 and
    Student.Dno=Department.Dno and
    Department.Dname='经济管理学院';
```

13. 答案

采用 SQL 语言编写一个嵌套查询:查询选修课程总学分在五个学分以上的学生的姓名。

```
select Sname
from Student
where Sno in(
    select Sno from SC,Course where
        SC.Cno=Course.Cno
    group by Sno
    having SUM(Ccredit)>=5);
```

14. 答案

采用 SQL 语言编写一个嵌套查询:查询各门课程的最高成绩的学生姓名及其成绩。

```
select Cno,Sname,Grade
from Student,SC SCX
where Student.Sno=SCX.Sno and SCX.Grade in
(
    select max(Grade)
    from SC SCY
    where SCX.Cno=SCY.Cno
    group by Cno
);
```

15. 答案

采用 SQL 语言查询所有选修了何小峰老师开设课程的学生姓名及其所在的院系名称。

```
select Sname,Dname
from Student,Department,SC
where Student.Sno=SC.Sno and Student.Dno=Department.Dno
and SC.Cno in(
select Cno
```

from TC
where TC. Tno=(select Tno from Teacher where Tname='何小峰')
);

16. 答案

采用 SQL 语言,在数据库中删除学号为 20091003085 的学生的所有信息(包括其选课记录)。

delete from SC where SC. Sno='20091003085';
delete from Student where Sno='20091003085';

17. 答案

采用 SQL 语言,将学号为 20091000863 的学生的"数据库原理"这门课的成绩修改为 80 分。

update SC set Grade=80
where Sno='20091000863'
and Cno=(select Cno from Course where Cname='数据库原理');

视图的创建与使用参考答案

1. 答案

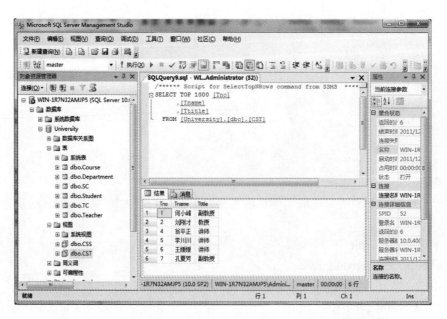

图 2-2　在 Microsoft SQL Server Management Studio 中创建视图

2. 答案
create view CSS as
select Sno,Sname,Ssex,Sage
from Student
where Dno=
(
select Dno from Department where Dname='计算机学院'
);
insert into CSS values('20101000911','尚方杵','M',16);
delete from CSS where Sno='20101000911';
drop view CSS;

3. 答案
create view CST as
select Tno,Tname,Ttitle
from Teacher
where Dno=
(
select Dno from Department where Dname='计算机学院'
);

4. 答案
select Sname from CSS where Sno in
(select Sno from SC SCX where SCX.Cno in(
select distinct SCY.Cno from SC SCY,TC
where SCY.Cno=TC.Cno and TC.Tno in
(select Tno from CST)));

5. 答案
delete from CST where Tno=1;
delete 语句与 reference 约束"FK__TC__Tno__1920BF5C"冲突。

数据库安全性参考答案

1. 答案
create login L1 with password='geoinfo';
create user U1 for login L1;
create login L2 with password='geoinfo';

create user U2 for login L2;
create login L3 with password='geoinfo';
create user U3 for login L3;
create login L4 with password='geoinfo';
create user U4 for login L4;
create login L5 with password='geoinfo';
create user U5 for login L5;
create login L6 with password='geoinfo';
create user U6 for login L6;
create login L7 with password='geoinfo';
create user U7 for login L7;
grant connect to U1,U2,U3,U4,U5,U6,U7;

2. 答案
grant select on Student to U1;

3. 答案
grant all privileges on Student to U2,U3;
grant all privileges on Course to U2,U3;

4. 答案
grant select on SC to public;

5. 答案
grant select,update on SC to U4;

6. 答案
grant insert on SC to U5 with grant option;

7. 答案
grant insert on SC to U6;

8. 答案
U6 不能对 U7 进行转授权。

9. 答案
revoke update on SC from U4;

10. 答案
revoke select on SC from public;

数据库完整性参考答案

1. 答案

(a)在 UNIVERSITY 数据库中创建表 STU_T,该表与 Student 表具有相同字段,其主码为 Sno。

创建 STU_T 表时,定义实体完整性(列级实体完整性);然后,删除 STU_T。

```
create table STU_T(
    Sno char(11)primary key(Sno),
    Sname char(8),
    Ssex char(2),
    Sage int,
    Dno int,
    );
drop table STU_T;
```

(b)创建 STU_T 表时,定义实体完整性(表级实体完整性);然后,删除 STU_T。

```
create table STU_T(
    Sno char(11),
    Sname char(8),
    Ssex char(2),
    Sage int,
    Dno int,
    primary key(Sno),
    );
    drop table STU_T;
```

(c)创建 STU_T 表后再定义其实体完整性 PK_SNO(提示:采用 alter 命令添加实体完整性)。

```
create table STU_T(
    Sno char(11)not null,
    Sname char(8),
    Ssex char(2),
    Sage int,
    Dno int,
    );
alter table STU_T add constraint PK_SNO primary key(Sno);
```

(d)删除数据表 STU_T 中的实体完整性 PK_SNO,然后,删除 STU_T。

```
alter table STU_T drop constraint PK_SNO;
drop table   STU_T;
```

2. 答案

(a)在 UNIVERSITY 数据库中创建表 SC_T,该表与 SC 表具有相同字段,其主码为 Sno、Cno。创建 SC_T 表时,定义实体完整性(表级实体完整性);然后,删除 SC_T。

```
create table SC_T(
    Sno char(11),
    Cno int,
    Grade int,
    primary key(Sno,Cno),
    );
drop table SC_T;
```

(b)创建 SC_T 表后再定义其实体完整性 PK_SC(提示:采用 alter 命令添加实体完整性);然后,删除数据表 SC_T 中的实体完整性 PK_SC,并删除 SC_T。

```
create table SC_T(
    Sno char(11)not null,
    Cno int not null,
    Grade int,
    );
alter table SC_T add constraint PK_SC primary key(Sno,Cno);
alter table SC_T drop constraint PK_SC;
drop table SC_T;
```

3. 答案

(a)在 UNIVERSITY 数据库中创建表 TC_T,该表与 TC 表具有相同字段,其主码为 Cno、Tno。

创建 TC_T 表时,定义实体完整性和参照完整性(采用表级实体完整性和参照完整性,被参照的数据表分别为 Teacher 表中的 Tno 和 Course 表中的 Cno);然后,删除 TC_T 表。

```
create table TC_T(
    Tno int,
    Cno int,
    Site char(50),
```

        primary key(Tno,Cno),
        foreign key(Tno)references Teacher(Tno),
        foreign key(Cno)references Course(Cno)
        );
    drop table TC_T;

(b)创建 TC_T 表后再定义其实体完整性和参照完整性(提示:采用 alter 命令增加实体完整性和参照完整性,被参照的数据表分别为 Teacher 表中的 Tno 和 Course 表中的 Cno)。

create table TC_T(
    Tno int not null,
    Cnoint not null,
    Site char(50),
    );
alter table TC_T add constraint PK_TC primary key(Tno,Cno);
alter table TC_T add constraint FK_TNO foreign key(Tno)references Teacher(Tno);
alter table TC_T add constraint FK_CNO foreign key(Cno)references Course(Cno);
drop table TC_T;

4. 答案

(a)在 UNIVERSITY 数据库中创建表 DEP_T,该表与 Department 表具有相同字段,其主码为 Dno。

创建 DEP_T 表时,定义实体完整性(列级实体完整性),并定义 Dname 唯一并且非空,Daddress 的默认值为"北一楼";最后,删除 DEP_T 的默认值"北一楼";最后,删除 DEP_T。

create table DEP_T(
    Dno int primary key(Dno),
    Dname char(50)unique not null,
    Daddress char(50)default '北一楼'
    );
  drop table DEP_T;

(b)创建 DEP_T 表时,定义实体完整性(表级实体完整性),并限制 Dno 的取值范围为00~99(提示:使用 check 进行约束);然后,删除 DEP_T。

create table DEP_T(
    Dno int,
    Dname char(50),
    Daddress char(50),

```
    primary key(Dno),
    check(Dno>0 and Dno<100)
    );
drop table DEP_T;
```

触发器参考答案

1.答案

在 UNIVERSITY 数据库中创建表 SGA_T,记录每个学生的所有选修课程的平均成绩,其主要字段包括 Sno、AverageGrade,其主码为 Sno。SGA_T 表的创建与初始化 SQL 语句如下:

```
create table SGA_T(
Sno char(11)primary key,
AverageGrade float);
insert into SGA_T
select Sno,avg(Grade)
from SC group by Sno;
```

(a)在 SC 表上定义一个 update 触发器,当修改某个学生的某一门选修课程的成绩后,自动重新计算所有的平均成绩,并更新到 SGA_T 表中。

第一步,对于触发器的创建可以采用界面进行操作,也可编写 SQL 语句进行操作。两者在本质上其实是一样的,这里采用界面操纵。在 SC 表节点下触发器上点击右键弹出菜单,选择新建触发器,生成初始化代码。

第二步,选择"查询→指定模板参数的值",触发器名称 Trigger_Name 的值指定为 SC_UPDATE_TRIGGER,触发器表 Table_Name 的值指定为 SC,数据变动类型 Data_Modification_Statements 的值指定为 UPDATE(图 2-3)。点击"确定",生成如图 2-4 中所示的初始化代码。

图 2-3 指定模板参数的值

```
-- examples of different Trigger statements.
--
-- This block of comments will not be included in
-- the definition of the function.
-- =============================================
SET ANSI_NULLS ON
GO
SET QUOTED_IDENTIFIER ON
GO
-- =============================================
-- Author:      Name
-- Create date:
-- Description:
-- =============================================
CREATE TRIGGER .SC_UPDATE_TRIGGER
    ON   .SC
    AFTER UPDATE
AS
BEGIN
    -- SET NOCOUNT ON added to prevent extra result sets from
    -- interfering with SELECT statements.
    SET NOCOUNT ON;

    -- Insert statements for trigger here

END
GO
```

图 2-4　生成的触发器初始化代码

第三步,在 BEGIN 和 END 之间编写处理代码,实现修改后自动更新 SGA_T 中的平均成绩数据。

delete from SGA_T;
insert into SGA_T
　　select Sno,avg(Grade)
　　from SC
　　group by Sno;

(b)在 SC 表上定义一个 insert 触发器,当添加某个学生的某一门选修课程的成绩时,自动重新计算所有学生的平均成绩,并更新到 SGA_T 表中。

create trigger SC_INSERT_TRIGGER
　　on SC
　　after insert
as
begin
　　set nocount on;
　　delete from SGA_T;
　　insert into SGA_T

```
        select SNO,avg(Grade)
        from SC
        group by Sno;
    end
```

(c)在 SC 表上定义一个 delete 触发器,当删除某个学生的某一门选修课程的成绩时,自动重新计算所有学生平均成绩,并更新到 SGA_T 表中。

```
create trigger SC_DELETE_TRIGGER
    on   SC
    after delete
as
begin
    set nocount on;
    delete from SGA_T;
    insert into SGA_T
        select SNO,avg(Grade)
        from SC
        group by Sno;
    end
```

(d)删除上面三个触发器,并删除数据表 SGA_T。

```
drop trigger SC_INSERT_TRIGGER;
drop trigger SC_UPDATE_TRIGGER;
drop trigger SC_DELETE_TRIGGER;
drop table SGA_T;
```

2.答案

在 UNIVERSITY 数据库中,我们规定每个老师至少必须上一门课程,最多只能同时上三门课程。

(a)在 TC 表上定义一个 insert 触发器,当添加某个老师新上的一门课程时,先检查该教师所上的课程总数是否达到了三门,如果达到三门的上限,则提示不能再为该教师新增课程。

```
create trigger TC_INSERT_TRIGGER
    on tc
    instead of insert
as
    declare @TNO int;
begin
```

```
    set nocount on;
    select @tno=Tno from inserted;
    if((select count(*) from TC where Tno=@TNO)>=3)
        print 'NO ACTION';
    else
    begin
        insert into TC
        select *
        from inserted;
    end;
end
```

(b) 在 TC 表上定义一个 DELETE 触发器,当删除某个老师所上的某一门课程时,先检查该教师所上的课程总数是否只有一门,如果是,则提示不能再为该教师减少课程。

```
alter trigger TC_DELETE_TRIGGER
    on   TC
    instead of delete
as
declare @TNO int,@CNO int;
begin
    set nocount on;

    select @TNO=Tno from deleted;
select @CNO=Cno from deleted;
    if((select count(*) from TC where Tno=@TNO)<=1)
    begin
        rollback;
        raiserror('NO ACTION',16,1);
    end;
    else
        delete from TC where Tno=@TNO and Cno=@CNO;
end
```

(c) 删除上面定义的两个触发器。

```
drop trigger TC_INSERT_TRIGGER;
drop trigger TC_DELETE_TRIGGER;
```

3.答案

在 UNIVERSITY 数据库中,我们规定每个老师只能属于一个系,在一个老师从一个系调动到另外一个系之前,需要先检查接受调动的系的编号是否合法,请在 Teacher 表上设计一个 before update 触发器来实现上述功能,然后删除该触发器。

```sql
create trigger TEACHER_TRIGGER
    on TEACHER
    instead of update
as
declare @DNO int,@TNO int;
begin
    -- SET NOCOUNT ON added to prevent extra result sets from
    -- interfering with select statements.
    set nocount on;

    -- Insert statements for trigger here
    select @DNO=Dno from inserted;
  select @TNO=Tno from inserted;
    if((select count(*)from Department where Dno=@DNO)<=0)
        raiserror( 'NO ACTION',16,1);
    else
    begin
    update Teacher
    set Dno=@DNO
    where Tno=@TNO;
    end;
end
```

数据库设计参考答案

请设计一个大学教学信息管理应用数据库 UNIVERSITY。其中,一个教师属于一个系,一个系有多名教师,每个系都有自己的办公地点。每个教师可以讲授多门课程,每个学生属于一个系,可以选修多门课程。每门课程具有一定学分,并可能有先导课程。

1.数据库概念结构设计

识别出教师(Teacher)、系(Department)、课程(Course)、学生(Student)四个实体。每个实体的属性和码如下:

(1)系 Department:系的编号 Dno、系的名称 Dname、系所在的办公地址 Daddress。主码为系的编号 Dno。

（2）学生 Student：学生学号 Sno、学生姓名 Sname、学生性别 Ssex、学生年龄 Sage、学生所属系编号 Dno。主码为学生学号 Sno。

（3）教师 Teacher：教师编号 Tno、教师姓名 Tname、教师职称 Ttitle。主码为教师编号 Tno。

（4）课程 Course：课程编号 Cno、课程名称 Cname、先导课程编号 Cpno、课程学分 Ccredit。主码为课程编号 Cno。

根据实际语义，分析实体之间的联系，确定实体之间一对一、一对多和多对多联系，绘制 E-R 图（图 2-5）。

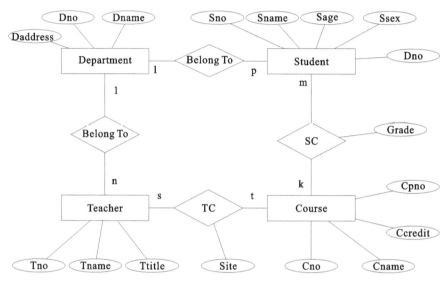

图 2-5 示例数据库 E-R 图

2. 数据库逻辑结构设计

按照数据库设计中概念结构向逻辑结构转换规则，根据所绘制 E-R 图，设计 UNIVERSITY 数据库逻辑结构（列出所有关系，确定每个字段的类型、长度等信息，以表的形式列出，画出数据库模式图），并写出相关 SQL 语句。

第一步，写出 UNIVERSITY 关系数据库模式。

（1）系的信息表 Department(Dno,Dname,Daddress)；
（2）学生信息表 Student(Sno,Sname,Ssex,Sage,Dno)；
（3）教师信息表 Teacher(Tno,Tname,Ttitle,Dno)；
（4）课程信息表 Course(Cno,Cname,Cpno,Ccredit)；
（5）学生选课表 SC(Sno,Cno,Grade)；
（6）教师授课表 TC(Tno,Cno,Site)。

第二步，结合选定的数据库管理系统（SQL Server），列出每个关系中每个属性的类型、长度等信息。

表 2-1 Department 关系属性表

关系名称		Department	关系别名			系的信息	
属性名	别名	类型	长度	值域	唯一	可空	备注
Dno	系编号	INT			Y	N	
Dname	系名称	CHAR	50		Y	N	
Daddress	系地址	CHAR	50		Y	Y	

表 2-2 Student 关系属性表

关系名称		Student	关系别名			学生信息	
属性名	别名	类型	长度	值域	唯一	可空	备注
Sno	学号	CHAR	11		Y	N	
Sname	姓名	CHAR	8		N	N	
Ssex	性别	CHAR	2	M/F	N	Y	
Sage	年龄	INT			N	Y	
Dno	系编号	INT			Y	Y	所在系

表 2-3 Teacher 关系属性表

关系名称		Teacher	关系别名			教师信息	
属性名	别名	类型	长度	值域	唯一	可空	备注
Tno	工号	INT			Y	N	
Tname	姓名	CHAR	50		N	N	
Ttitle	性别	CHAR	50		N	Y	
Dno	系编号	INT			Y	Y	所在系

表 2-4 Course 关系属性表

关系名称		Course	关系别名			课程信息	
属性名	别名	类型	长度	值域	唯一	可空	备注
Cno	课程号	INT			Y	N	
Cname	课程名	CHAR	50		Y	N	
Cpno	先导课	INT			N	Y	
Ccredit	学分	INT			Y	Y	所在系

表 2-5 SC 关系属性表

关系名称	Course		关系别名			课程信息	
属性名	别名	类型	长度	值域	唯一	可空	备注
Cno	课程号	INT			Y	N	
Sno	学号	CHAR	11		Y	N	
Grade	成绩	INT			N	Y	

表 2-6 TC 关系属性表

关系名称	Course		关系别名			课程信息	
属性名	别名	类型	长度	值域	唯一	可空	备注
Cno	课程号	INT			Y	N	
Tno	工号	INT			Y	N	
Site	位置	CHAR	50		N	Y	

第三步,画出数据库模式图,如图 2-6 所示。

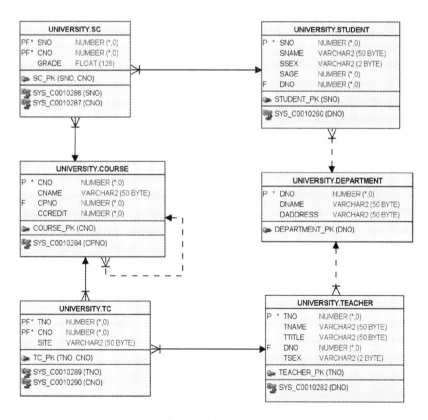

图 2-6 示例数据库结构

第四步，设计用户子模式。例如，设计计算机系学生信息视图CSS。这里没有需要设计的其他用户子模式，故可以直接进入下一步物理设计。

3. 数据库物理结构设计

数据库物理结构设计首先根据数据库逻辑结构自动转换生成，然后再根据实际应用需求，设计数据库的索引与存储结构。这里主要给出数据库结构的SQL语言实现。

```
create table Department(
    Dno int,
    Dname char(50),
    Daddress char(50),
    primary key(Dno)
);
create table Student(
    Sno char(11),
    Sname varchar(8),
    Ssex varchar(2),
    Sage int,
    Dno int,
    primary key(Sno),
    foreign key(Dno)references Department(Dno)
);
create table Teacher(
    Tno int primary key,
    Tname char(50),
    Ttitle char(50),
    Dno int,
    foreign key(Dno)references Department(Dno)
);
create table Course(
    Cno int primary key,
    Cname char(50),
    Cpno int,
    CCredit int,
    foreign key(Cpno)references Course(Cno)
);
create table SC(
    Sno char(11),
    Cno int,
```

```
        Grade int,
        primary key(Sno,,Cno),
        foreign key(Sno)references Student(Sno),
        foreign key(Cno)references Course(Cno)
);
create table TC(
        Tno int,
        Cno int,
        Site char(50),
        primary key(Tno,Cno),
        foreign key(Tno)references Teacher(Tno),
        foreign key(Cno)references Course(Cno)
);
```

存储过程与函数参考答案

答案

(1)定义一个无参数存储过程 DecreaseGrade,更新所有学生成绩,将其降低 5%,并调用该存储过程。

```
--创建名为 DecreaseGrade 的无参数存储过程
create procedure DecreaseGrade
as
begin
    update SC set Grade=Grade * 0.95;
end
    select Grade from SC

exec DecreaseGrade
```

(2)定义一个带输入参数存储过程 IncreaseGrade,将课程号为 1 的所有学生成绩提升 5%;要求课程号作为存储过程参数传入,并调用该存储过程。

```
--创建名为 IncreaseGrade 的有输入参数的存储过程
create procedure IncreaseGrade
    @ccno int
as
begin
```

```
            update SC set Grade=Grade * 1.05 where Cno=ccno;
end
select Grade from SC
```

--执行名为 IncreaseGrade 的有输入参数的存储过程(传入参数)
```
exec IncreaseGrade '1'
```

（3）定义一个带有输入和输出参数的存储过程 AverageStudentGrade，计算一个学生的所有选修课程的平均成绩，要求学号作为输入参数，计算结果——该生的所有选修课平均成绩作为输出参数；调用该存储过程，并输出计算结果。

--创建名为 AverageStudentGrade 的有输入参数和输出参数的存储过程
```
create procedure AverageStudentGrade
@paramsno nvarchar(50)
as
    begin
        select sg. ag
        from(select Sno s,avg(Grade)ag
            from SC group by Sno)sg
                where sg. s=@paramsno;
    end
```

--执行名为 AverageStudentGrade 的有输入参数和输出参数的存储过程
```
exec AverageStudentGrade '20091000863'
```

（4）删除存储过程 IncreaseGrade、DecreaseGrades。

```
drop procedure IncreaseGrade;
drop procedure DecreaseGrades;
```

（5）定义一个带有输入参数的自定义函数 CalculateAverageStudentGrade，计算一个学生的所有选修课程的平均成绩，要求学号作为输入参数，返回该生的所有选修课平均成绩；调用函数，并输出计算结果。

--创建名为 CalculateAverageStudentGrade 的有输入参数的函数
```
create function CalculateAverageStudentGrade
(
@paramsno nvarchar(50)
)
returns float
```

```
as
   begin
   return
   (
       select sg.ag
       from(select Sno s,avg(Grade)ag
           from SC group by Sno)sg
               where sg.s=@paramsno
   )
   end
```

select dbo.CalculateAverageStudentGrade('20091000863')

(6)删除函数 CalculateAverageStudentGrade。

drop function CalculateAverageStudentGrade

(7)定义一个存储过程,采用普通无参游标实现计算学校开设的所有课程的学分之和。

```
create procedure pro_7
as
declare @ss int

declare @cc int

declare cur cursor for select Ccredit from Course
set @ss=0
   open cur
   fetch next from cur into @cc
   while @@fetch_status=0
     begin
     set @ss=@ss+@cc
     fetch next from cur into @cc
     end
   close cur
   select @ss
   deallocate cur

exec pro_7
```

(8) 定义一个存储过程,采用 REF CURSOR 实现计算学校所有学生选修课程的成绩之和。

```sql
create procedure pro_7
as
declare @ss int
set @ss=0
declare @cc int
declare cur cursor scroll dynamic optimistic
for select Grade from SC
   open cur
   fetch next from cur into @cc
   while @@fetch_status=0
     begin
       set @ss=@ss+@cc
       fetch next from cur into @cc
     end
   close cur
   select @ss
deallocate cur

exec pro_7
```

(9) 定义一个存储过程,采用带参数游标实现按照学号计算学生的平均成绩。

```sql
create  proc  pro_9
(
  @sno_invarchar(50)

)as
declare @g int;
declare @paramsno  varchar(50);
set @paramsno=@sno_in;
declare cursor1 cursor
for select sg.ag
        from(select Sno s,avg(Grade)ag from SC group by Sno)sg
        where sg.s=@paramsno;

begin
    open cursor1;
```

```
    fetch next from cursor1 into @g;
    while(@@FETCH_STATUS=0)
    begin
        print(@g);
         fetch next from cursor1 into @g;
    end;
    close cursor1;
    deallocate cursor1;

end

execute pro_9 '20091000961'
```

数据库应用开发(C++)参考答案

1. 答案

配置 SQL Server 数据库的 ODBC 连接,采用 C++开发一个基于 MFC 和 ODBC 的数据库应用程序,实现以下功能:输出所有学生选修课程名称和成绩。

(a)配置 University 数据源。

第一步:打开 Microsoft ODBC Administrator(图 2-7)。

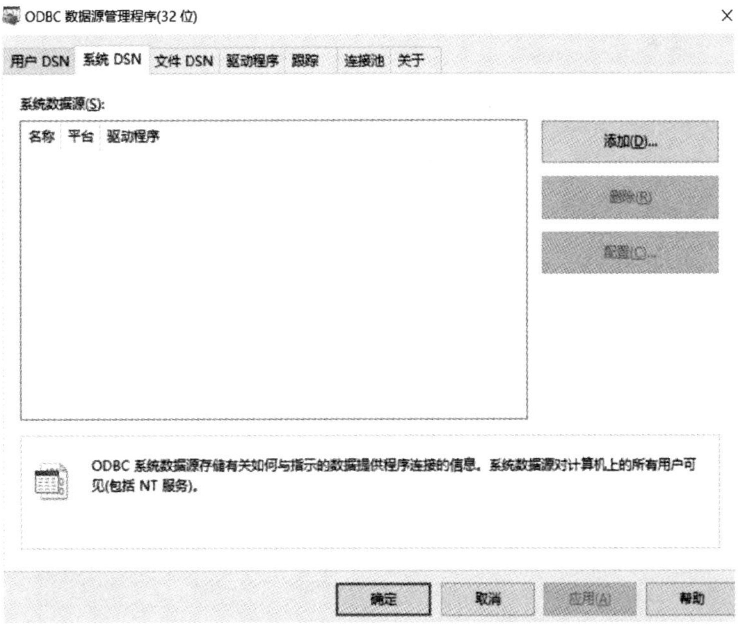

图 2-7　ODBC 系统 DSN

第二步:选择"系统 DSN",选择"Add..."按钮,选择"SQL Server"驱动程序创建数据源(图 2-8)。

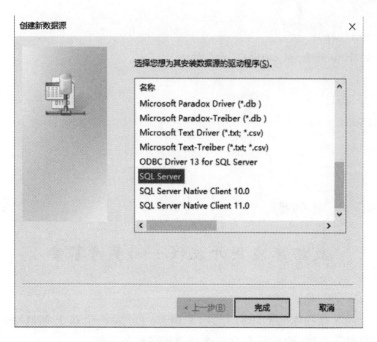

图 2-8 数据源的 SQL Server 驱动

第三步:配置 ODBC 驱动与数据源,使用 windowsNT 验证,更改默认数据库,然后点击"完成",并测试数据源(图 2-9~图 2-14)。

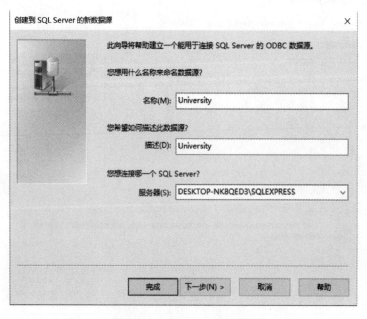

图 2-9 创建到 SQL Server 的新数据源名称

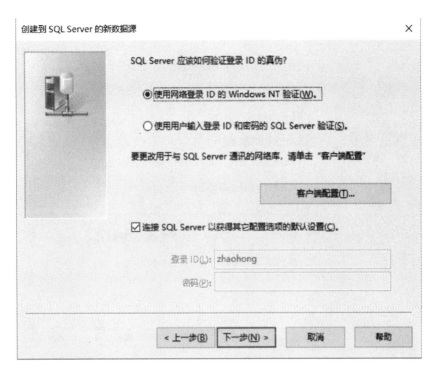

图 2-10　创建到 SQL Server 的新数据源的用户信息

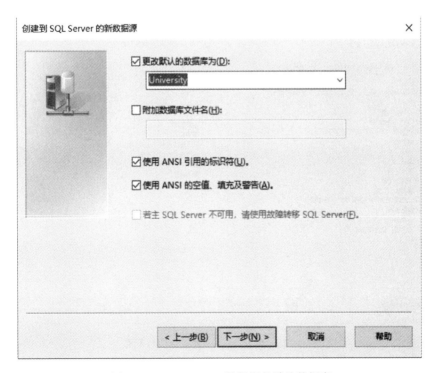

图 2-11　SQL Server 数据源的默认数据库

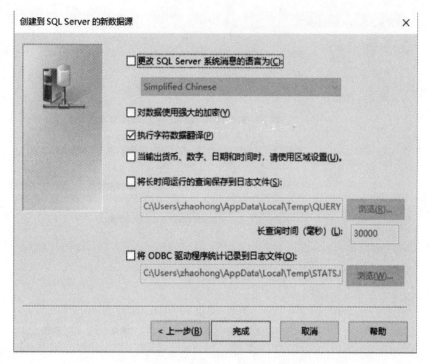

图 2-12　SQL Server 数据源的附属信息

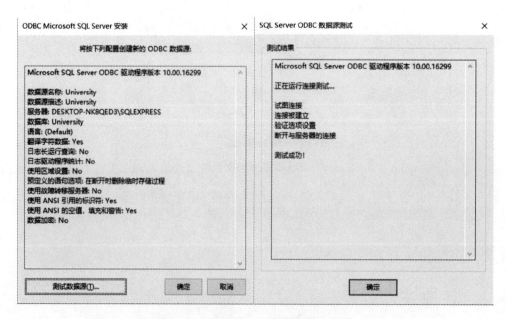

图 2-13　SQL Server 数据源信息汇总

图 2-14 基于 SQL Server 驱动的 University 数据源

(b) 基于 MFC ODBC 开发数据库应用程序。由于直接采用 C 语言版的 ODBC API 编程比较繁琐，所以较多的数据库组件或库都对 ODBC 的 C 语言版本 API 进行了面向对象的封装。其中，MFC 中的 CDatabase、CRecordset 等类封装 C 语言版的 ODBC API 的大部分功能。为简便起见，我们这里直接采用 Visual Studio 2010 进行基于 ODBC 的数据库应用程序开发。

第一步，按照图 2-15～图 2-18 的方式生成示例工程。

(a) 选择"MFC 应用程序"

图 2-15 示例工程模板

(b)选择"基于对话框(D)"程序类型

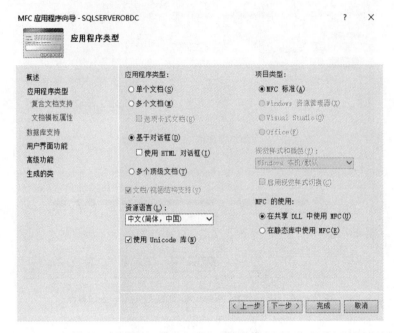

图2-16 基于对话框的程序模板

然后一直点"下一步"至完成即可。

第二步,在对话框中添加list控件用于显示查询的数据,更改为report显示风格,如图2-17,并添加变量(图2-18)。

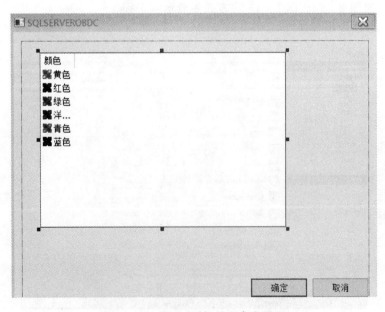

图2-17 基于对话框的程序主界面

图 2-18 为对话框添加变量

第三步,在 stdafx.h 文件中添加 afxdb.h 头文件。

```
#include <afxcontrolbars.h>      // 功能区和控件条的 MFC 支持
#include <afxdb.h>
```

第四步,在对话框类 CSQLSERVEROBDCDlg 中添加 CDatabase 成员变量 m_database 和相应函数声明。

```
    inline CDatabase * GetDatabase()
    {
        return &m_database;
    }
    BOOL OpenDatabase();
    int ReadRecords();
public:
    CListCtrl m_list_teacher;
    CDatabase m_database;
};
```

第五步,编码实现 OpenDatabase() 函数,主要是调用 CDatabase 连接并打开数据库。

```
BOOL CSQLSERVEROBDCDlg::OpenDatabase() {
    CString strDBSource = _T("UNIVERSITY_PDBORCL");
    CString strUserName = _T("UNIVERSITY");
    CString strPassword = _T("cug");
    CString strConnect;
    strConnect.Format(
        _T("DSN=%s;UID=%s;PWD=%s"),
        strDBSource,
        strUserName,
        strPassword);

    TRY {
        m_database.OpenEx(strConnect, CDatabase::noOdbcDialog);
    }
    CATCH (CDBException, ex1)   {
        AfxMessageBox(ex1->m_strError);
        AfxMessageBox(ex1->m_strStateNativeOrigin);
        return FALSE;
    }
    AND_CATCH (CMemoryException, ex2)   {
```

代码如下：

```
BOOL CSQLSERVEROBDCDlg::OpenDatabase(){
CString strDBSource=_T("UNIVERSITY");
CString strUserName=_T("sa");
CString strPassword=_T("123456");
CString strConnect;
strConnect.Format(
    _T("DSN=%s;UID=%s;PWD=%s"),
    strDBSource,
    strUserName,
    strPassword);

TRY {
    m_database.OpenEx(strConnect,CDatabase::noOdbcDialog);
}
CATCH(CDBException,ex1){
    AfxMessageBox(ex1->m_strError);
    AfxMessageBox(ex1->m_strStateNativeOrigin);
    return FALSE;
}
AND_CATCH(CMemoryException,ex2){
    ex2->ReportError();
```

```
        AfxMessageBox(_T("Memory Exception"));
        return FALSE;
    }
    AND_CATCH(CException,ex3){
        TCHAR szError[256];
        ex3->GetErrorMessage(szError,256);
        AfxMessageBox(szError);
        return FALSE;
    }
    END_CATCH
    return TRUE;
}
```

第六步,编码实现 ReadRecords()函数,实现从数据库中查询返回记录,写入 list 控件显示。

```
int CSQLSERVEROBDCD1g::ReadRecords() {
    CString strSQL =
        _T("SELECT * FROM Teacher");
    int rc = -1;
    int n=0;
    TRY{
        CRecordset record(GetDatabase());
        record.Open(AFX_DB_USE_DEFAULT_TYPE, strSQL, 0);
        rc = record.GetRecordCount();
        CODBCFieldInfo f1, f2;
        record.GetODBCFieldInfo((short)0, f1);
        record.GetODBCFieldInfo((short)1, f2);

        CString Tname, Ttitle;
        int Tno, Dno;
        CDBVariant var;
        while (!record.IsEOF()) {
            record.GetFieldValue((short)0, var);
            if(var.m_dwType!=DBVT_NULL)
                Tno=var.m_iVal;
            var.Clear();
            record.GetFieldValue((short)1, Tname);
            record.GetFieldValue((short)2, Ttitle);
            record.GetFieldValue((short)3, var);
            if(var.m_dwType!=DBVT_NULL)
                Dno=var.m_iVal;
            var.Clear();

            record.MoveNext();
```

```
            //写入list控件
            CString temp;
            temp.Format(_T("%d"),Tno);
            m_list_teacher.InsertItem(n, temp);
            m_list_teacher.SetItemText(n, 1, Tname);
            m_list_teacher.SetItemText(n, 2, Ttitle);
            temp.Format(_T("%d"),Dno);
            m_list_teacher.SetItemText(n, 3, temp);
            n++;

        }
    }
    CATCH(CException, e) {
        TCHAR szError[100];
        e->GetErrorMessage(szError, 100);
        AfxMessageBox(szError);
    }
    END_CATCH
    return rc;
}
```

具体代码如下：

```
int CSQLSERVEROBDCDlg::ReadRecords(){
    CString strSQL=
        _T("SELECT * FROM Teacher");
    int rc=-1;
    int n=0;
    TRY{
        CRecordset record(GetDatabase());
        record.Open(AFX_DB_USE_DEFAULT_TYPE,strSQL,0);
        rc=record.GetRecordCount();
        CODBCFieldInfo f1,f2;
        record.GetODBCFieldInfo((short)0,f1);
        record.GetODBCFieldInfo((short)1,f2);

        CString Tname,Ttitle;
        int Tno,Dno;
        CDBVariant var;
        while(!record.IsEOF()){
            record.GetFieldValue((short)0,var);
            if(var.m_dwType!=DBVT_NULL)
```

```
                Tno=var.m_iVal;
            var.Clear();
            record.GetFieldValue((short)1,Tname);
            record.GetFieldValue((short)2,Ttitle);
            record.GetFieldValue((short)3,var);
            if(var.m_dwType!=DBVT_NULL)
                Dno=var.m_iVal;
            var.Clear();

            record.MoveNext();


            CString temp;
            temp.Format(_T("%d"),Tno);
            m_list_teacher.InsertItem(n,temp);
            m_list_teacher.SetItemText(n,1,Tname);
            m_list_teacher.SetItemText(n,2,Ttitle);
              temp.Format(_T("%d"),Dno);
            m_list_teacher.SetItemText(n,3,temp);
              n++ ;

        }
    }
    CATCH(CException,e){
        TCHAR szError[100];
        e->GetErrorMessage(szError,100);
        AfxMessageBox(szError);
    }
    END_CATCH
    return rc;
}
```

第七步，在 OnInitDialog() 函数的最后返回之前调用，实现 list 控件的初始化，用 OpenDatabase() 函数连接数据库和 Read Recordset() 函数获取查询结果。

```
DWORD dwStyle = m_list_teacher.GetExtendedStyle();
dwStyle |= LVS_EX_FULLROWSELECT;
dwStyle |= LVS_EX_GRIDLINES;
m_list_teacher.SetExtendedStyle(dwStyle);
//教师数据表的list控件初始化
m_list_teacher.InsertColumn(0, _T("Tno"), LVCFMT_CENTER, 25);
m_list_teacher.InsertColumn(1, _T("Tname"), LVCFMT_CENTER, 80);
m_list_teacher.InsertColumn(2, _T("Ttitle"), LVCFMT_CENTER, 80);
m_list_teacher.InsertColumn(3, _T("Dno"), LVCFMT_CENTER, 80);


if (OpenDatabase()) {
    ReadRecords();
    //TODO:输出列表中的查询结果
    return TRUE;
}
else {
    return FALSE;
}
```

第八步,运行结果如图 2-19 所示。

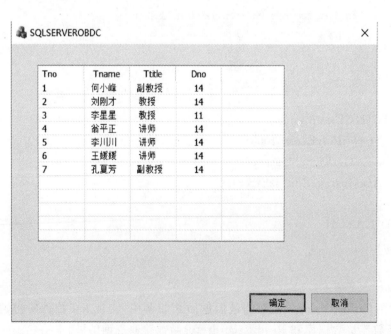

图 2-19 程序运行结果

数据库应用开发(Java)参考答案

1. 答案

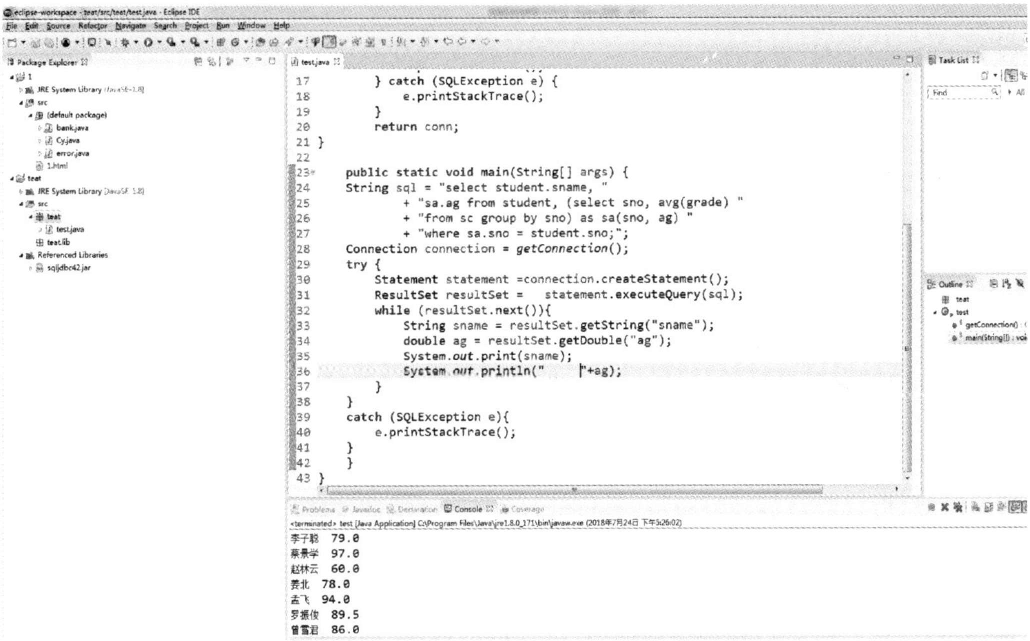

图 2-20 Eclipse 连接 Microsoft SQL Server

2. 答案

```java
package teat;

import java.sql.*;

public class test{

    public static Connection getConnection(){
        Connection conn=null;
        try {
            Class.forName("com.microsoft.sqlserver.jdbc.SQLServerDriver");//找到 oracle 驱动器所在的类
            String url="jdbc:sqlserver://127.0.0.1:1433;DatabaseName=University";//URL 地址
            String username="sa";
            String password="061144";
```

```
            conn=DriverManager.getConnection(url,username,password);
        }catch(ClassNotFoundException e){
            e.printStackTrace();
        }catch(SQLException e){
            e.printStackTrace();
        }
            return conn;
    }

    public static void main(String[] args){
    Stringsql="select student.sname,"
        +"sa.ag from student,(select sno,avg(grade)"
        +"from sc group by sno)as sa(sno,ag)"
        +"where sa.sno=student.sno;";
    Connection connection=getConnection();
    try{
        Statement statement=connection.createStatement();
        ResultSet resultSet=statement.executeQuery(sql);
        while(resultSet.next()){
            String sname=resultSet.getString("sname");
            double ag=resultSet.getDouble("ag");
            System.out.print(sname);
            System.out.println("     "+ag);
        }
    }
    catch(SQLException e){
    e.printStackTrace();
    }
    }
}
```

数据库应用开发(C#)参考答案

采用C#开发一个基于ODP.NET的数据库应用程序,实现以下功能:输出每个学生选修课程平均成绩。

(1)构建C#程序,选择Windows Form App(.NET Framework),如图2-21所示。

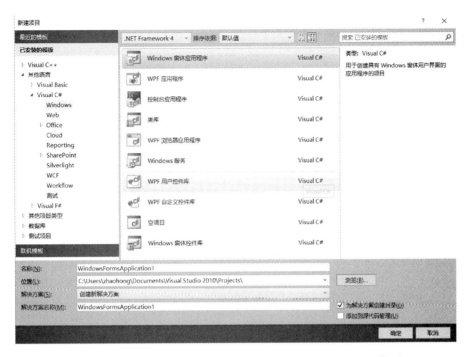

图 2-21 选择 Windows Form App(.NET Framework)工程类型

(2)在 Form 中放置 DataGridView 和 Button 两个控件,如图 2-22 所示。

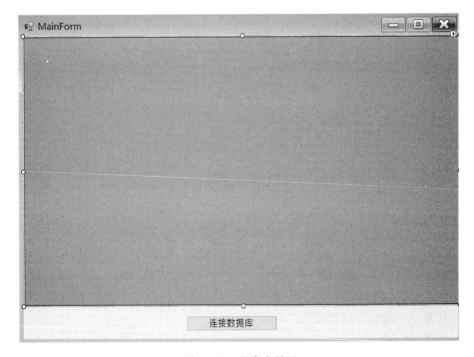

图 2-22 程序主界面

(3) 在 Button 的单击响应事件中连接数据库,实现对数据库的查询。

```csharp
private void button1_Click(object sender, EventArgs e)
{
    String connsql = "server=.;database=University;integrated security=SSPI"; // 数据库i
    SqlConnection con = new SqlConnection(connsql);
    string sql = "select sno, avg(grade) sg from sc group by sno;";
    SqlCommand com = new SqlCommand(sql, con);
    try
    {
        con.Open();
        MessageBox.Show("成功连接数据库");
        SqlDataAdapter myda = new SqlDataAdapter(sql, con); // 实例化适配器

        DataTable dt = new DataTable(); // 实例化数据表
        myda.Fill(dt); // 保存数据

        dataGridView1.DataSource = dt; // 设置到DataGridView中




    }

        dataGridView1.DataSource = dt; // 设置到DataGridView中



    }
    catch (Exception ex)
    {
        MessageBox.Show("错误信息:" + ex.Message, "出现错误");
    }
    finally
    {
        con.Close();
        MessageBox.Show("成功关闭数据库连接");
    }

}
}
```

(4) 此时可能无法连接到 Sql Server 数据库,需要执行如下步骤。

(a) 开始菜单→所有程序→Microsoft SQL Server→配置工具→SQL Server 配置管理器→网络配置→MSSQLSERVER→双击"TCP/IP"→协议→已启用→选"是"。

SQL Server 配置管理器→网络配置→MSSQLSERVER→双击"TCP/IP"→IP 地址→IPAll→TCP 端口→输入"1433",点击"确定"。

(b) 开始菜单→所有程序→Microsoft SQL Server→配置工具→SQL Server 配置管理器→SQL Server 服务→ SQL Server(MSSQLSERVER2008)→右键重新启动。

在命令行下输入 netstat-an,如果找到有"0.0.0.0:1433",就说明 SqlServer 在监听了。
(c)操作系统→安全中心→Windows 防火墙→ 例外→添加程序。
C:\ProgramFiles\MicrosoftSQLServer\MSSQL10.MSSQLSERVER\MSSQL\Binn\sqlservr.exe

运行结果如图 2-23 所示。

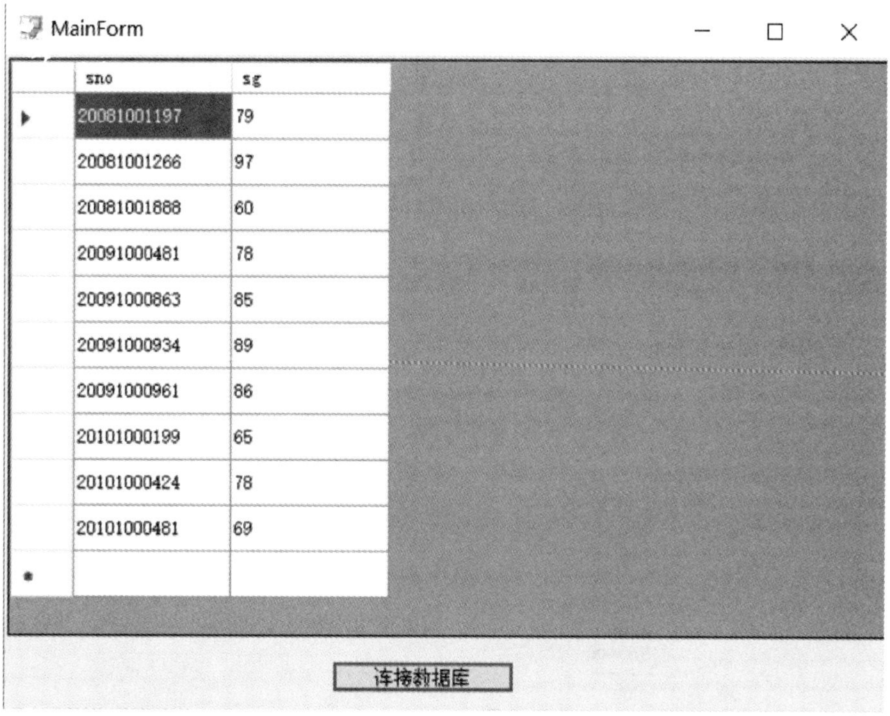

图 2-23 运行结果

数据备份与恢复参考答案

1.答案

逻辑备份(导出)UNIVERSITY 数据库全部内容到指定文件中(图 2-24)。备份整个数据库,恢复时恢复所有。优点是简单,缺点是数据量太大且非常耗时。步骤为:右键数据库→任务→备份(图 2-25)。

SQL 语句如下:
BACKUP DATABASE [University] TO DISK=N'C:\Program Files\Microsoft SQL Server\MSSQL14.SQLSERVER2017\MSSQL\Backup\University.bak' WITH NOFORMAT,NOINIT, NAME=N'University-完整 数据库 备份',SKIP,NOREWIND,NOUNLOAD,STATS=10
GO

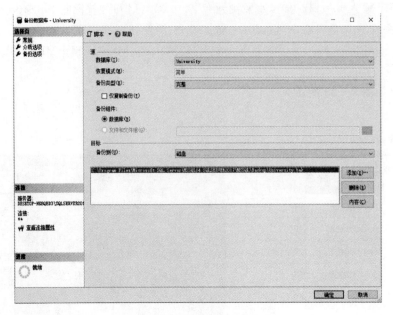

图 2-24 选择备份文件

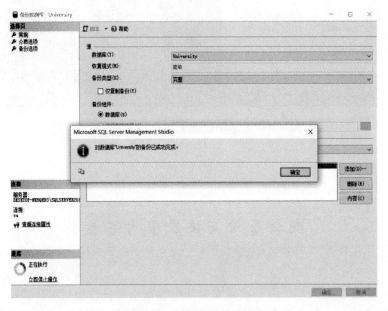

图 2-25 备份过程完成

2. 答案

增量数据备份(Differential Backups)：所谓增量，就是以某个起始时间点的全量数据为基础，备份该时间点以后的数据。而起始时间点的全量数据，就是通过全量备份而来的。如果有人告诉你"每周一进行全量备份，每天进行一次增量备份"，这就意味着，星期一作一次全量备份，形成一个起始时间点的全量数据；星期二备份星期一以来的数据；星期三也备份星期一以来的数据……星期天也备份星期一以来的数据。到第二周的星期一时，又执行一次全量备份，再开始新的备份周期。如果要恢复星期三的数据,则要先恢复星期一的全量数据,然后再恢复在星期一到星期三之间的增量数据(图 2-26、图 2-27)。

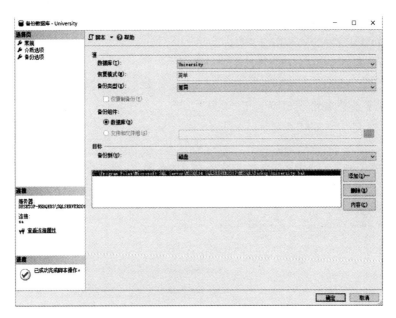

图 2-26 差异备份

图 2-27 数据库恢复

SQL 语句如下:

BACKUP DATABASE [University] TO DISK=N'C:\Program Files\Microsoft SQL Server\MSSQL14. SQLSERVER2017\MSSQL\Backup\University. bak' WITH DIFFERENTIAL,NOFORMAT,NOINIT, NAME=N'University -完整 数 据 库 备 份',SKIP, NOREWIND,NOUNLOAD, STATS=10

GO

主要区别在于参数备份类型变为了"差异"。

3. 答案

逻辑恢复(导入)UNIVERSITY 数据库全部内容(图 2-28)。

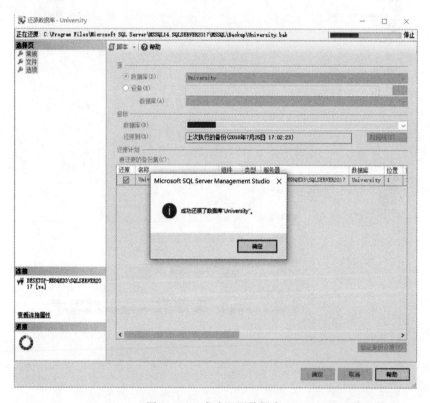

图 2-28 成功还原数据库

SQL 语句如下:
USE [master]
RESTORE DATABASE [University] FROM DISK=N'C:\Program Files\Microsoft SQL Server\MSSQL14. SQLSERVER2017\MSSQL\Backup\University. bak' WITH FILE =1, NOUNLOAD, STATS=5
GO

4. 答案

(1) 部分备份 SQL 语句。

Backing Up Specific Files or Filegroups
BACKUP DATABASE { database_name | @database_name_var }
 <file_or_filegroup> [,...n]
  TO <backup_device> [,...n]
  [<MIRROR TO clause>] [next-mirror-to]
  [WITH { DIFFERENTIAL | <general_WITH_options> [,...n] }]
[;]

(2) 部分还原 SQL 语句

-- To perform the first step of the initial restore sequence
-- of a piecemeal restore：
RESTORE DATABASE { database_name | @database_name_var }
   <filcs_or_filegroups> [,...n]
[FROM <backup_device> [,...n]]
  WITH
    PARTIAL,NORECOVERY
     [,<general_WITH_options> [,...n]
     |,\<point_in_time_WITH_options—RESTORE_DATABASE>
    ] [,...n]
[;]

-- To Restore Specific Files or Filegroups：
RESTORE DATABASE { database_name | @database_name_var }
   <file_or_filegroup> [,...n]
[FROM <backup_device> [,...n]]
  WITH
  {
     [RECOVERY | NORECOVERY]
     [,<general_WITH_options> [,...n]]
  } [,...n]
[;]

(3) 参数部分。{ database_name | @database_name_var } 是备份事务日志、部分数据库或完整的数据库时所用的源数据库＊。如果作为变量(＊@database_name_var)提供,则可以将此名称指定为字符串常量(@database_name_var=＊＊＊database name)或指定为字符串数据类型(ntext 或 text 数据类型除外)的变量。

<file_or_filegroup> [,...n]只能与 BACKUP DATABASE 一起使用,用于指定某个数据库文件或文件组包含在文件备份中,或指定某个只读文件或文件组包含在部分备份中。

FILE={ logical_file_name | @logical_file_name_var }是文件或变量的逻辑名称。

FILEGROUP={ logical_filegroup_name | @logical_filegroup_name_var }是文件组或变量的逻辑名称。在简单恢复模式下,只允许对只读文件组执行文件组备份。

综合实验(课程设计)参考答案(略)

主要参考文献

王珊,萨师煊.数据库系统概论[M].5版.北京:高等教育出版社,2014.

王珊,张俊.数据库系统概论(第5版)习题解析与实验指导[M].北京:高等教育出版社,2015.

刘福江,张传维,等.数据库课程设计与开发实操[M].北京:科学出版社,2017.

何珍文.空间数据库实验教程[M].武汉:中国地质大学出版社,2019.